A Golden Love Story

A Golden Love Story

Phyllis Seymour
with foreword by Sir Colin Cowdrey C.B.E.

The Lutterworth Press
Cambridge

The Lutterworth Press
P.O. Box 60
Cambridge
CB1 2NT

First published in Great Britain by The Lutterworth Press, 1993
Copyright © Phyllis Seymour, 1993

British Library Cataloguing in Publication Data:
A catalogue record is available for this book from the British Library

All rights reserved.
No part of this publication may be reproduced,
stored in a retrieval system, or transmitted in any form
or by any means, electronic, mechanical, photocopying,
recording or otherwise, without the prior permission
in writing of the publisher.

Printed in Great Britain by
Redwood Books, Trowbridge, Wiltshire

Contents

Foreword by Sir Colin Cowdrey 1

Chapter 1
A Case of Puppy Love 3

Chapter 2
A Change in the Season 9

Chapter 3
Sally 15

Chapter 4
Dreams of a White Christmas 20

Chapter 5
Spring by the Sea 26

Chapter 6
Goodbye, Salvador! 31

Chapter 7
Puppies' Progress 35

Chapter 8
Grown-up Goings on 40

Chapter 9
Sibling Rivalry and All That 45

Chapter 10
Sally in the Wars 49

Chapter 11
Canadian Caper 53

Chapter 12
Memories of Sam 59

Chapter 13
Right to the Heart of it 63

Chapter 14
And Then There Were None 69

Chapter 15
The Last Goodbye 73

For Douglas, who made it all possible.

'The origin of the Golden Retriever is less obscure than most of the Retriever varieties, as the breed was definitely started by the first Lord Tweedmouth last century, as shown in his carefully kept private stud book and notes, first brought to light by his great-nephew, the Earl of Ilchester, in 1952.

In 1868 Lord Tweedmouth mated a yellow wavy-coated Retriever (Nous) he had bought from a cobbler in Brighton (bred by Lord Chichester) to a Tweed Water-Spaniel (Belle) from Ladykirk on the Tweed. These Tweed Water-Spaniels, rare except in the Border Country, are described by authorities of the time as like a small Retriever, liver-coloured and curly-coated. Lord Tweedmouth methodically line-bred down from this mating between 1868 and 1890, using another Tweed Water-Spaniel and outcrosses of two black Retrievers, an Irish Setter and a sandy-coloured Bloodhound. (It is now known that one of the most influential kennels in the first part of the century which lies behind all present day Golden Retrievers was founded on stock bred by Lord Tweedmouth.)'

USEFUL ADDRESSES

Beckenham Dog Training Club
Chief Instructor
Mrs Christine Cornish
120 Burns Avenue
Sidcup
Kent

PDSA Head Office
PDSA House
Whitechapel Way
Priorslee
Telford
Shropshire

My thanks go to Sylvia Knight for her invaluable help in typing and editing my manuscript - PS.

Our Golden Family Tree

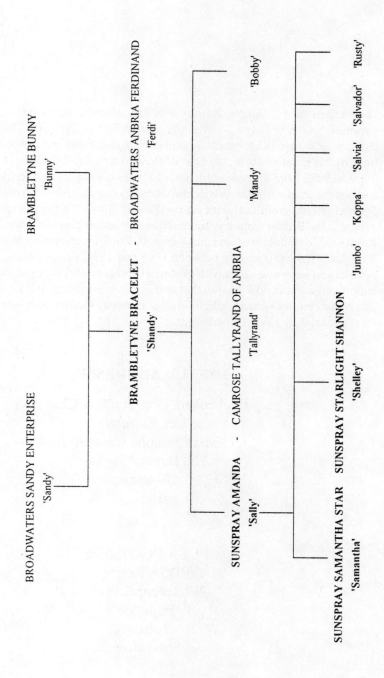

FOREWORD
by Sir Colin Cowdrey CBE

When I was invited to write a foreword to this book, my memory turned back to my childhood and the dogs my parents owned at that time. There was a terrier called Patch who would play football when required, and Darkie, owned by my grandmother, who was self-appointed guard of my pram when the family came home on leave from India. Sadly, in later years, due to my almost constant travels, I had to forego the pleasure of having my own dog. I missed having a friend whom I could walk and feed daily, and who would be my constant companion.

The author of this story has indeed been fortunate to have known the company, love and trust of her canine family, and has set out to portray this, together with her own feelings for them, and that of her family.

Any dog owner will recognise the fun, fulfilment and ultimate heartbreak associated with the love of these special friends. I commend to you this moving Golden Love Story.

Colin Cowdrey

*"To all beings and all things
we shall be as relatives."*

SIOUX INDIANS

Chapter 1
A CASE OF PUPPY LOVE

Our story begins in the late Summer of 1964. I knew then that I'd be leaving my job with Barclays Bank and embarking on a whole new way of life the following January. There'd be a new house, a new name and a new venture: becoming a dog owner, or - as it would turn out in practice - being owned by one myself! Just one, at least to start with . . .

My father had always wanted to buy me a dog when I was old enough to look after one. My first experience of puppies was at the age of eight when I was photographed with a litter of black labradors and I promised myself then that one day I would have just such a beautiful creature of my own.

But, sadly, it was not to be. My father died when I was 10 and there could be no more thought of puppies for the moment. In fact, the idea had to be dismissed completely two years later when my mother sold our house and moved into a flat - and a flat, as everyone knows, is no place for a big dog, the kind I'd always hoped to have.

* * * *

Though I'd never owned one, I'd always admired and loved dogs. My fiancé, Douglas, however, had grown up with dogs. His mother kept Cairn terriers when he was a boy and more recently he'd owned a very autocratic but highly amusing Pekinese which he'd bought for his own children. Douglas and I had always agreed that we would have a dog and that, when our dog finally arrived, he or she would be a large one. More than that we hade no idea which breed we fancied. However, in September, 1964, we made one of our many trips to Eastbourne to visit Douglas's parents and an aunt of mine, who also lived there. We always travelled via Godstone in Surrey, where we liked to stop at a delightful hotel, which also combined an antique shop and a small tea-room. That afternoon in the tea-room was to settle it. On this occasion, we were greeted by two lovely golden retrievers. 'I like that dog,' said Douglas, when the first 'golden' appeared. So did I and, from then on, there was never a second thought: the breed had 'spoken' to us there and then.

* * * *

The owner of the hotel was quite naturally delighted when we took so quickly to her pets and gave us the name of their breeder, a Mrs Gwen Medhurst of Cudham. We duly contacted Mrs Medhurst, who asked us to come for an interview where we explained that we very much wanted a puppy to take to our new home in January, if at all possible.

A Golden Love Story

Mrs Medhurst knew she would have no puppies in January, as none of her bitches had been in season, nor would be until the following spring. But she was more than willing to enquire among her 'golden' breeder friends to see what could be done.

Her final question was 'dog or bitch?' and I was frankly at a loss when Douglas insisted the choice must be mine and mine alone. I finally found myself saying: 'I think I would like a bitch,' and so the matter was settled. We left Mrs Medhurst with a promise that she would telephone as soon as she had any news of a litter.

It was the week before Christmas when the promised call came. A litter of 'goldens' had arrived on December 4 in Aylesbury to a bitch called Brambletyne Bunny. The breeders were Mr and Mrs Tripptree but there was no question of us choosing a puppy from the litter - they were all booked. All, that is, except for one bitch who, by the end of that momentous phone call, was well and truly booked herself!

The pups, seven in all, would be ready to leave for their new homes eight weeks after the birth. Their new owners were treated, quite rightly, on a first come, first served basis. And so, as we were the last of the new owners, there was no choice of colour or size: we simply had the last girl puppy there was. (I prefer to call them girls rather than bitches). I couldn't have known it then, but she was to be the best Christmas news and the most wonderful present ever. I called her Shandy.

* * * *

Shandy arrived in our lives on Sunday, 31 January 1965, the day before my birthday. We'd moved into Elmwood Cottage, our new home, just over a week before during a heavy fall of snow. Douglas's company, which had moved us, had thoughtfully laid white carpets along the stairs and landing so you can probably imagine the muddy horrors that lay in store for us already!

Joining us in our new home was my brother, Alfred, some 20 years my senior and a bachelor. I'd given my mother a promise before she died five years earlier that I'd look after Alfred, who had suffered from 'nerves' and a speech defect since childhood, and Douglas generously supported me in my promise by welcoming him to Elmwood Cottage.

Alfred's job was in Sydenham, just a few miles from the house and Douglas was director and company secretary of a large retail store in Lewisham known then as Chiesmans and now part of the House of Fraser group. The fourth member of our new household was Violet Williamson. She had come to work for me in my single days, following my mother's death, at the large, lofty Victorian flat which I had shared with my mother and Alfred since my father's death in 1942.

The flat was far too large for me to keep really clean the way I liked. I was working very long hours at first and I then joined the bank where my hours could be most erratic, so Violet's help had been invaluable.

Violet came to work for us every Monday, Wednesday and Friday morning and, as always, she worked like a Trojan the day we moved into Elmwood. She didn't 'live in' but was always willing to stay, if needed.

Elmwood is a country-style house, in a London suburb, with a wealth of oak beams, oak floors and a large stone fireplace in the lounge, and a smaller replica in the hall. I was often asked if we had bought the dog to go with the house or the other way around. Certainly, the one complemented the other perfectly.

* * * *

On the afternoon of 31 January, I set off for Cudham with Douglas and Alfred to collect Shandy. Mrs Medhurst had picked her up on the previous day and, when we knocked on her door at 3 pm, she appeared with a small furry bundle in her arms. The puppy was dark golden in colour, with a lovely head and dark eyes like clear pools.

'Is that our Shandy?' I asked tentatively.

'Yes,' said Mrs Medhurst, 'she's tired, we've been playing.'

We sat and chatted for a while and watched Shandy at play. There was the payment to be made and the diet sheet to be explained. This was entirely new to me. I was completely clueless on all aspects of dog owning and puppy rearing and, needless to say, I was to make all my mistakes with Shandy, many of which she would greatly enjoy!

One was to take her with me when I went to our local newsagents. Each time a great deal of fuss was made of her and a delicious, fattening bar of chocolate produced. In fact, she became well known all along that shopping parade. One assistant at the health food shop took a great deal of interest in Shandy, admiring her appealing face and soft, friendly nature, and it was to this lady that I went for sound advice and help some years later when Shandy developed arthritis.

* * * *

The new addition to our family had travelled well from Cudham to Sydenham, only to be sick just as we pulled into the drive. Since she hadn't been fed before the journey, this presented no big problem, and our first duty once indoors was to give her a dish of warm milk. Never have I seen anything disappear so quickly! Douglas was quite horrified and pleaded with her to take her time when offering her a second dish. But this, of course, fell on deaf ears, and where food was concerned, they remained deaf to all admonishments for the rest of her life.

We followed the golden rule of putting Shandy out after drinking, eating or sleeping to allow nature to take its course, showering her with praise after each performance. That first evening Shandy was completely worn out after the upheaval of two strange houses within 24 hours and several equally strange people, and she curled up in front of the log fire and fell asleep as only a puppy can.

Although it was on, I hardly looked at the television all night. I couldn't take my eyes off the sleeping bundle of fur at my feet, already a fairly dark golden fur that had a silky feel to it, and those silky ears, even darker, giving a hint of the colour she'd eventually become.

Time for bed and mistake number one. We settled Shandy down after her supper in a large box covered with blankets in the warmest corner of the kitchen. Then we said 'goodnight', put out all the lights and went to bed. But we'd been there for barely half an hour when the wailing began. We waited to see if our new baby would settle down, but the pitiful cries went on unabated, and I finally gave in. Downstairs I was greeted by a frantic puppy who grabbed the bottom of my dressing gown and pulled me into the kitchen.

I sat with her for a while before coaxing her back into bed and making my way quietly back to my own, only to be called out again ten minutes later. This time I tried to sound very cross but all to no avail. The crying by now was desperate and I decided that, if anyone was to sleep at all that night, I should either take Shandy upstairs with us or stay down with her until she fell asleep and I could creep away.

Since Shandy was yet to be house-trained I adopted the latter plan, and the next night was the same story. The following day was a Tuesday and the vet was due to call to give Shandy a check-up and her first injection. I decided to mention the bedtime problem to him.

'Get an old clock, wind it up, put it in her bed with a warm hot water bottle wrapped in an old woollen garment of yours,' he said, 'and you will have no more trouble.' The puppy was missing the warmth and company of her brothers and sisters, and the ticking clock would sound like her mother's heartbeat, which would comfort her in the dark. There was no doubt in the vet's mind that this little trick would work, but I remained sceptical!

However, we duly produced a clock that night, along with a hot water bottle and an old woolly. Then we went to bed, hearts in our mouths, and waited. To our great surprise, there wasn't a single sound that night, or any other night from then on.

The late Sam Costa had a radio programme about that time, which encouraged people to write in for advice from other listeners on all kinds of domestic problems. The 'what to do with a new puppy at night' question was quite a common one and I wrote in to report my success. My letter was duly read out on the air and subsequently a large number of people wrote to me to say how well the vet's advice had worked for them too.

* * * *

As the weeks went by, I could see that Shandy was quick to learn. Toilet training was hardly necessary, for instance. She would run to the door in the breakfast room leading to the garden after only about two weeks - she was not always in time, but the message had obviously registered in her brain. Of course, there were still occasional mistakes indoors but, as Douglas once remarked, her little accidents were only the size of a half crown in those far off pre-decimal days.

Shandy was growing fast and had arrived at the 'leggy' stage at around three to six months. We decided to take her back to see Mrs Medhurst and she confirmed what I'd already suspected, that she could do with being a bit fatter. I didn't realise at that time how rapid the growth cycle of a young pup really is and, being afraid to overdo the feeding, I'd been giving Shandy less than the desired amount of food. No wonder she was always ravenous.

Any vet will tell you that it's almost impossible to overfeed a rapidly growing dog. So Shandy's meals were stepped up as we followed the diet sheet more closely this time and she started to gain the required weight.

Mrs Medhurst called to see us about a month later, bringing Shandy's pedigree for me to sign and keep, and doubtless to see how she was progressing. She departed with a word of warning about the proximity of a major road and implored me never to venture from our house to the park gate, only yards away, without putting her on a lead. This was sound advice and in all her sixteen years we never did.

* * * *

A week after her injections were all completed, it was time to take Shandy out into the world. We'd got used to the collar and lead routine in the garden and around the house, so going out with them was no problem. Shandy seemed very keen to explore the big world outside the front door. When a car came by she simply sat down and watched it pass. But she was very shy when it came to strange people and other dogs.

I mentioned this to our vet who advised us to contact a dog training club: the Kennel Club would put us in touch with a reliable one in our area, he said. 'I strongly advise you to take her as soon as she is six months old,' he went on. 'It can do nothing but good.' How right he was!

Our first training evening in early June '65 turned out to be rather a traumatic one for little Shandy. All those people and all those dogs of all shapes and sizes! She promptly dived under a chair and spent most of the evening there.

'Don't worry,' the organisers said. 'Bring her again next week and start her in the beginners' class.' This we did and this time Shandy settled down and seemed anxious both to learn and to please. We started, as anyone who has been to dog training classes will know, with basic heel work on a lead, when dogs learn to sit and stay and to go down and stay on command.

The lesson lasted half an hour and we practised at home every day, along the landing for about five minutes, going over the exercises we'd learned on our first evening. Shandy was a quick student and soon she had mastered the recall and retrieving a dumb-bell. She did well in the beginners' section and, at the annual dance in November, when all the different awards were presented, I was amazed to be called up and presented with the 'Trophy for Progress'. The chairman smiled and said, as he shook my hand and made the presentations, 'You have done wonders with Shandy in just a few weeks.'

We continued enthusiastically with the training and I was delighted when we were promoted to the next class within a very short time, and then moved up to the top one in less than a year. After that, Shandy competed in the annual competition every year and she always came somewhere in the roll of honour, receiving all sorts of prize cards and medals. She no longer went to class in her very late years but, even so, she never forgot her training.

* * * *

Shandy's second treat, next to her meals, was riding in the car. She adored it and would leap in at every opportunity, sitting firmly in the front seat, which she claimed as her own. She was transferred there from the back seat as a puppy to try to avoid travel sickness and it was to remain 'her' seat for the rest of her life. She wouldn't budge from it, not even for me.

I would be obliged to sit in the back while Shandy - once she tired of gazing out of the window - would rest her head on the back of the seat and fix me with an adoring, mildly apologetic gaze. Our windows were always open, wound down all round to about four inches and, in more recent years, our sun roof would always be open in good weather. But never has any of our dogs been allowed to travel with its head hanging out of the window. This is a deplorable way for a dog to be allowed to travel and always makes me very annoyed. It is bad for their eyes and any number of dangerous situations can arise.

* * * *

Our first disaster involving food and Shandy came when she was still less than a year old. Douglas's sister and her husband came to stay one weekend and we'd planned to go to Tunbridge Wells for the day. The weather was warm and we decided it would be wiser to leave Shandy at home in the cool of the kitchen and breakfast room. We gave her a good run in the morning and my brother, Alfred, stayed in with her until lunch-time when he fed her and left her to sleep, or so he thought!

I'd bought a leg of lamb for the following day, when we were expecting friends for lunch. I'd left the meat on a plate on the table, intending to put it in the fridge, but went out and forgot all about it. When we came home at about 5 pm, there was Shandy, fast asleep on her back, tummy bulging, satisfied grin on her face. And there on the mat, at the door leading to the garden amid pieces of broken plate, were the remains of the meat, bearing no resemblance whatsoever to the leg of lamb. Realising I had less than half an hour before closing time, ghastly visions floated before me of our friends arriving to find no lunch waiting for them.

Desperately, I telephoned our butcher in the hope he might have something left, and he saved the day by offering to deliver another joint on his way home. Needless to say, Shandy quickly became a firm favourite with him. A few more like her and business would be well and truly booming!

Chapter 2
A CHANGE IN THE SEASON

Shandy's first season was nothing like as traumatic as we'd been led to expect. She made no attempt to get out without us then or at any other time in her entire life. I continued to exercise her in the local park, but never at peak times in the evening and mid-afternoon. I always tried to go very early or very late and keep to a remote part of the park. All went well, except for two incidents. The first occasion was when we were away on holiday.

We had left Shandy at home with Alfred and Violet, our help, who always lived in when Douglas and I were away. She would look after Shandy's meals while Alfred took care of the exercising. But, unfortunately, Shandy came into her first season just after we'd left on one such holiday and, one evening, when Alfred took her for a late walk, the inevitable happened.

Poor Alfred was very worried when he told Violet all about it and he got her to phone the vet straight away and ask for an injection for Shandy. He knew this had to be given with 24 hours of a mating to ensure no unwanted pups would be born. Violet had never heard of such a thing and wouldn't believe it until Alfred had shown her the chapter in our golden retriever book which dealt with season and mating. Once administered, however, the injection not only kills the sperm but keeps the bitch in season for a further closely guarded three weeks.

* * * *

Shandy was less than a year old when I made my first attempt at showing her. It was a big local show held at Beckenham Baths and neither she nor I had ever been in a show ring before and we hadn't a clue what to do or, more to the point, what not to do. We were soon to find out!

Our class was called and we entered the ring. 'Oh, this is easy!' I thought. 'We just watch what everyone else does and do the same.' But that was easier said than done.

What I didn't know at that stage was that the dog not only has to stand properly for the judge but must stand still while the judge looks at the teeth and jaws and feels the body all over even down to the legs and tail. Our Shandy was having none of this. She was just like a flea on the end of the lead. As to leads, there is a show lead made of nylon which all dogs wear when in the ring but no-one had told me this and Shandy went in wearing her choke chain and heavy leather lead. Let's just say that her placing in the show is best left to the imagination.

We tried again, this time in Bexley Heath and this time we had acquired the regulation lead and taught Shandy how to stand correctly. But, once again, Shandy refused to be handled by anyone outside our household and an uproar ensued as soon as the judge, John Norman, approached her. He tried again and again without success, with the same result as before. Two years later Mr Norman told me that he had liked Shandy as soon as he saw her in the ring. If only she had shown in the proper fashion that evening, she would have been placed first in a fairly large class, he said knowingly. What he did not know was that one of his own dogs, Broadwaters Sandy, was actually Shandy's father. So it was no wonder he'd taken such a shine to her.

We often wondered whether Shandy had inherited her sweet temperament from her dam, Brambletyne Bunny, or her sire, Broadwaters Sandy. Perhaps it was both or even neither. But one thing we did know for certain, it couldn't have been better. I never knew her to snap at us or anyone else or for that matter at any other dog, despite provocation from people and other pets in the park. Shandy simply never lost her cool. This was why I was at a loss to understand her reluctance to be handled in the ring by a stranger. I began making enquiries among my new friends at training classes. They wisely suggested I should persevere and wait until she'd passed her second season, when they felt sure she'd calm down and start to mature.

How right they were. At her third show in aid of the PDSA and presided over by a well-known judge, Shandy won two first prizes. She went on to win many more prizes between first and fifth for several years and came full circle in her show career by winning first place in the same hall at the same show for the PDSA in the veteran class for dogs at the age of fourteen.

What had impressed the judge were her lovely white teeth, all still intact at that great age. It was Alfred who took Shandy into the ring on that occasion. I had by now, taken over the running of the show, and for obvious reasons I made it a rule never to take a dog of mine into the ring at any show I'd organised.

Life for Shandy followed an even and fairly uneventful pattern for the first three years: A run in the park twice a day, training classes every Tuesday evening and as many car rides as we could manage to please her.

It is to my regret that she never came on holiday with us or stayed away with us at any time; except of course to our home on the coast, which we always think of as our second home and therefore not a holiday home in that sense. But my joy, on looking back, is that she never slept anywhere but in her own home, except for one occasion after an operation at fourteen years of age, when the vet kept her in overnight to keep an eye on her.

* * * *

We decided we'd have Shandy mated on her third heat. 'It would be nice for her to have a litter. She is sure to be a good mother with such a sweet temperament,' we said.

The first step, of course, was to find a nice mate for her. We decided to contact Mr Norman in Barnet, the judge who had admired Shandy at the show in Bexley Heath. He introduced us to his stud dog, Broadwaters Anbria Ferdinand.

Meeting Ferdinand and his relatives was quite an experience for us. We arrived, as arranged, one evening at Mr Norman's bungalow to be greeted by his wife and no less than 14 golden retrievers! They all lived indoors during the day, bitches in one room, dogs in another. At bedtime, Mr Norman simply said one word: 'Bed!' and all the 'goldens' filed through the hall in orderly fashion to their respective kennels. It is a sight I shall never forget. Neither will I forget Shandy's own mating.

* * * *

It was the tenth day of her heat when Ferdinand was brought in and introduced. He was, of course, very interested in Shandy, but she was not in the least bit interested in him. We tried and tried to get them together, but it was no use and, in the end, poor old Ferdinand was so fed up he just turned and walked out of the room. 'Oh no, don't you start,' cried an exasperated Mr Norman. But that was that as far as the spurned bridegroom was concerned. Mr Norman suggested that we bring Shandy back in two or three days time. We took her back on the twelfth day and a successful mating did at last take place. Douglas later described our experience with Shandy in a letter to his father. In his reply, Mr Seymour Senior said he found it all most interesting but could not understand the reluctance of the stud dog.

* * * *

The following nine weeks passed all too quickly, as life went on for us all in the normal way. We stepped up the amount of food we gave the mother-to-be during the final four weeks and added supplements, such as calcium and various vitamin-packed products. We watched vainly for signs of an enlarged tummy or waistline, but none appeared.

Shandy took great care of herself as the date of the birth approached and her pace around the park became a good deal slower. One afternoon, on one of our quiet walks, a large German shepherd bounded around a corner, intent on having a game, and knocked Shandy flat. She picked herself up, without any apparent harm, but I felt very concerned, with the birth only days away.

I slept downstairs with her for several nights, in case the pups arrived early. On the evening of Friday 25 May, Shandy refused her dinner and, when I phoned the vet, he said the birth should take place at any time within the next 24 hours. Shandy went into labour at around noon the next day and the pups, four in all, were born without any fuss or bother in a large whelping box in the breakfast room.

The room is well away from the main part of the house, with a door leading into the garden, and it is convenient in every way for a nursing doggy mother, as it is both quiet and private. I was advised by the vet that the more privacy and peace Shandy had the better during the first two weeks. As soon as the first pup had arrived, I phoned Douglas at the office to tell him everything was going well since he was never home before 6 pm on Saturdays and would miss it all. He did in fact manage to get home early, just in time to see the last two puppies born.

It was an experience I shall never forget - so unlike a human birth, when the baby is taken away and washed by a nurse or midwife before being presented to its mother. Each pup arrives in what looks like a thick cellophane bag. Immediately, the mother rips this open and begins to lick the pup all over to get it clean and livened up. If no human help is on hand, the pup will crawl to its mother and sniff around till it finds a teat. Even so a helping hand from his 'other' mum, in this case me, is gratefully received.

A Golden Love Story

All appeared well, as the new family settled down for the night, and we gave Shandy a well deserved drink of warm milk laced with a generous dash of Brandy. Never have I seen a drink disappear with such speed! Then she had to be coaxed away from the pups long enough to do what was necessary in the garden before the entire household could go to bed. A good mother will always need coaxing at first to leave her puppies, even for a second. We got to bed around 11 pm and I could hardly wait for the morning so that I could go down to see how our little family was progressing. All appeared well. The proud mother greeted us with her eyes and her tail waved like a windmill in triumph. The day turned out to be a busy one, with friends ringing to enquire after the furry family and, of course, wanting to come round for a peep.

Our vet had advised us not to let anyone unconnected with the household in to see the puppies for the first two weeks. Unfortunately, I'd promised Violet's grandchildren a look as early as possible. When they arrived on the second day, they were so reluctant to leave that the little boy screamed for all he was worth as he was led away, which, I am sure, was the cause of Shandy's nervousness later in the day.

* * * *

We went to bed that night, convinced all was well again. But, on going down to the breakfast room early next morning, Douglas found one puppy lying all quiet and motionless on his own, well away from his mother and the rest of the family. Douglas called frantically upstairs to me to come and look at the puppy. 'I think it's dead.' He shouted. Before he'd even finished speaking, Shandy had picked up the pup, ran as hard as she could upstairs to my bedroom and laid the stricken little animal on the bed in my lap. Then she returned quickly to her other pups. As far as she was concerned, she'd done the wisest thing for us all, and now it was up to me.

It was 7 am on Bank Holiday Monday when we telephoned the vet and he was with us within half an hour. He examined the little mite and announced that he had pneumonia. After giving him an injection, he suggested that I put him somewhere warm, but I was prepared to sit and nurse him myself all day, if necessary. Sadly, I was to nurse him for only two hours before his little heart gave up.

The vet returned to take him away and I have always regretted not keeping him at home and burying him in the garden. The rest of the day seemed somehow unreal as it always does when someone close is taken away. The little pup had only been a matter of 48 hours old, but even so was already part of the family. Shandy, for her part, decided wisely that life must go on and her remaining babies lacked for nothing. Her routine of feeding, washing and general nursing continued. She was a wonderful and patient mother, but she was not above administering a 'clip' when needed to a youngster who'd stepped out of line. Shandy was, of course, still having all her supplements with her food: she had loved eating all her life and that particular trait was now standing her in good stead.

* * * *

The pups were growing apace now: two light-coloured ones (a boy and a girl) and a dark golden pup, just like mum. We had decided to keep a puppy before Shandy was mated and, as soon as the litter arrived, we picked out the palest pup, as a complete contrast to Shandy. The name Sally was quickly chosen for our new little girl and, for the whole of her life, I felt a bit like Gracie Fields, singing out her name to coax her in from the garden.

I was very reluctant to leave mother and babies alone for long and, even on a shopping trip around the local shops, was eager to get home and spend more time with our furry friends. This was our first ever litter, but I must say that the thrill and excitement of each new litter is just as great as the last. How many pups are there? In how many shades of gold? How many boys? How many girls? It's all such a miracle the way it happens.

We were lucky: We only had two prospective homes to worry about with our first litter and things couldn't have turned out any better. I was put in touch with one family in Hayes through my dog training club. They'd only recently lost their 'golden' and were keen to have another. They wanted the girl and an equally nice family in Beckenham were interested in the boy. Both families came to see the pups when they were just four weeks old and they were straightaway christened Mandy and Bobby. It was only a week before that I'd been feeling concerned about weaning them successfully. They both enjoyed the milk Shandy provided so much that I couldn't begin to imagine giving them meat, fish or anything else for that matter. But I was soon reassured by the book I was following at the time and by advice picked up among 'doggy' people at the club. The thing to do, I discovered, was to take a chunk of beef, scrape a sharp knife across it a couple of times, roll the thin meat left on the knife into a ball, open the tiny mouth and pop it in. 'Once tasted, it will never be refused,' said the book and, sure enough, after that first introduction to solid food, the pups never looked back.

Soon the infants were eating from their own bowls, mopping their meals up like sponges. Shandy was still feeding Sally, Mandy and Bobby and would gladly have continued, if I hadn't provided solid food. She loved feeding and washing them long after her maternal duties should have finished. I made a big mistake in letting her feed Sally until she was almost four months old. All very nice for Sally, but dear old Shandy was never to regain her nice light undercarriage again.

Apart from meat and soaked biscuit meal, the pups were now having goat's milk instead of cow's as it is far richer and more acceptable to them. Since that time, an excellent number of milk products for nursing mothers and puppy weaning have come on to the market and goat's milk is in short supply, most of it going to the big supermarket chains. The next four weeks passed all too quickly. The pups were changing in size, colour and character. Mandy was a very deep gold, like her mother, while Sally and Bobby were a mid-golden colour. They were fast becoming characters in their own right and were best of friends, despite a few healthy squabbles.

* * * *

It took six weeks before they were all on five meals a day, three solid and two milk. The day was fast approaching when Mandy and Bobby would be leaving for their new homes and, delighted as I was with their new owners, I wished the day of departure would never come: the pups were part of the family. But come it must, I realised, and

I was comforted by the knowledge that I'd found marvellous homes for the pair of them.

They left us at eight weeks and Bobby was the first to go. It was 7 pm on Sunday 23 July 1967, when the Simmonds family arrived with a relative who happened to be a vet. He'd come to look Bobby over and hopefully pass the little rascal fit, not that I had any doubts in that direction. All the surviving pups were as tough as old boots!

The farewell scene didn't last long as Bobby was whisked away, with a final hug and kiss from me. I couldn't help feeling sad as they all drove off although Bobby would only be a ten minute car ride away in Beckenham and it was agreed that I could visit him when he'd settled into his new home. I promised to phone the next day to see how Bobby's first night had gone. When I closed the door behind them and settled down to play with the remaining pups the next morning seemed a long way off. But when I did make the call, Clare Simmonds assured me that everything was fine. I'd given her the tip about the ticking clock and hot water bottle but, as things turned out, they weren't needed. I felt happy and well satisfied that everything had gone so nicely for young Bobby and that his new family were so delighted with him.

* * * *

Mandy stayed with us for another two and a half weeks before her new owners could collect her and it was the evening of Wednesday 9 August when Milly Osmond and her family came to pick Mandy up. Again, I felt a great sense of loss as I hugged and kissed her before handing her over. Milly could see I was close to tears, and was quick to comfort me.

'You can rest assured that she'll be well loved,' she said, as she cradled Mandy in her arms. And so she was, for over sixteen wonderful years, as in his turn was Bobby, for the fifteen years of his life.

Mandy would also be close by, only a mile or two away in Hayes, and the following day Milly rang to report that all was well.

I called on the pups after two weeks, to see how they were getting along. Bobby was having a high old time, chasing around his new house and garden, but he greeted me at length during a brief break from his play. He and his family were happy and that was all that mattered to me.

When I arrived at Mandy's house, Milly opened the door and put her finger to her lips. We tiptoed into the lounge where Mandy was sound asleep on a rug in front of the fireplace, and sat down as quietly as possible, speaking in a whisper. It took about five minutes for Mandy to wake up and, as she looked around, she suddenly realised I was there. She went wild with excitement and for the rest of her life, whenever I appeared, I would receive this same greeting.

She adored Milly and was her dog in every sense, but she never forgot me. All our pups greeted me happily whenever I went to visit them or they were brought over to see me but, in Mandy's case, there was always that extra special joy whenever we met.

Chapter 3
SALLY

If our tiny Sally missed her brother and sister, she certainly didn't show it and Shandy, of course, was delighted with the new arrangement. Her maternal duties were to continue for several more months and, as far as face and ear washing were concerned, went on for the rest of their lives, with never a protest from Sally.

For the first few months, Sally slept with her mother in the large Goddard dog bed we had at the time. This worked well until Sally grew even bigger than her mother. Try as they would to exist comfortably together, it just wasn't working out. Long-suffering Shandy would fix us with a glance that said: 'No room!' so we gracefully gave way and bought them a second bed.

It had been three years since Shandy's arrival as an eight-week old pup and I had forgotten, even in that short time, how playful and demanding a small bundle of fur can be. But I hadn't forgotten the mistakes we'd made during Shandy's formative years - like allowing her titbits all too frequently and all those chocolate bars from our nice newspaper man she'd happily gobbled up.

We'd learned our lesson from errors like these and never again had to cope with a weight problem, except on the odd occasion when we would return from holiday, having left a doting Violet in charge. On our return, waistlines would be closely inspected and biscuits and biscuit meal cut down. Exercise would be stepped up and, of course, training practice and classes resumed, all of which resolved the problem fairly quickly.

* * * *

Having mother on hand proved to be a big bonus in many ways for young Sally. It helped when visitors called and when we took trips out in the street or by car, and it calmed her nerves at injection time. Consequently, she had a good deal of confidence and was much less nervous than Shandy had been at the same age. We discovered that, when a puppy stays with her mother, she tends to mature more slowly and, when in the company of other dogs, appears to have stayed younger and more alert.

I always think it a pity more people don't or can't keep more than one dog, though I must stress most strongly that one's circumstances should always be seriously considered when taking on a dog at all. When Sally was around four months old, she made her debut at the dog club and she was full of confidence, unlike mother on her first appearance, taking a great interest in all that went on. All pups, when they appear in the hall for the first time, cause a great flurry of excitement and, while most enjoy the adulation, for those with a shy nature, the experience can be a bit off-putting.

But it's amazing how quickly nine out of ten dogs react, overcoming any stage fright and quickly adapting to their new surroundings. For this reason, we always encouraged new handlers to attend classes with their young puppies, to get them used to the atmosphere and general hurly-burly of the club. As a rule, they are invited to sit and watch for around a month, before launching their beginners on to the scene.

I began to train Sally, little by little, along the landing at home, away from all the many distractions downstairs. Like Shandy, she was quick to learn and more than eager to please. Two months passed quickly and we were ready to go into the first class. We didn't have the usual pre-beginners' class, followed by a month's course doing very simple things. We were thrown in at the deep end, doing heel work, sits, stays and recall, all on a lead, of course.

In the meantime, our mother and daughter team was presenting a most attractive picture to the world - at home by the fire, walking along the road, running in the park, or out on the prom at the seaside - they were greatly admired wherever they went.

Sally was growing fast and becoming longer in the leg and back than her mother or sister. She was quick to learn and one of her many endearing little habits was to sit on the stairs and peer around the corner to see if I was watching her and what I was wearing.

Both dogs knew my walking clothes and would get very excited if they suddenly spotted me in these. They'd also get excited if they thought I'd gone upstairs to get ready for an outing but, if I reappeared in a dressing gown, they'd stop dead in their tracks and stare at me in disbelief. Sally left the guarding of the house to Shandy, who was very, very good at it. Well, why live with a capable 'doggy' relative and do the barking yourself?

We trained Sally to walk to the car as soon as it was safe for her to venture into the world outside, once her injections were all completed. It always angers me when owners take tiny puppies out, completely unprotected, into the street. Any vet will tell you that they can contract that old enemy, distemper, as well as many other diseases, including leptospirosis, commonly known as lamp-post disease. I always think that, if any of these thoughtless owners had to suffer from a dose of one of these unpleasant diseases themselves, they would think twice about allowing their pets out unprotected.

* * * *

Shandy always adored the car, but I felt that Sally was less keen, putting up with it simply to be with her mother and us on our many outings. These included frequent visits to Tamarisk, our home on the Kent coast all year round. As young dogs, they always loved the walk from the house to the front, taking us along the prom, with the beach on our left, and grassy slopes with beach huts on our right.

We'd walk and they'd run all the way, until the prom ran out and we found ourselves on flat, marshy land. Then we'd make our way back, at the same pace, the whole trip taking about an hour. If the weather was fine, we'd do the same thing in the afternoon or early evening, great exercise for all concerned! Sally, unlike her mother, never had a weight problem in all her long life, due partly, I like to think, to the lessons I'd learned with Shandy, but also to her own make-up and metabolism.

A new pattern had been established by the autumn of '67 and Shandy and Sally had become a pair of dogs who liked to do everything together. Together they quickly

Sally

adapted to living in two different places and being looked after, when we were away, by Alfred and Violet. This particular year I felt better about going on holiday and leaving Shandy in their care, now that she had such good company. Much as an animal may love her owner, or anyone else who cares for her, there's a lot to be said for a bit of company of her own kind!

* * * *

We set off for Santa Margarita in Italy for three weeks, taking our boat with us. We had a lovely trip, visiting many different resorts, including Portofino, and came home rested and refreshed after a hectic, demanding puppy-rearing summer.

Sally had grown a good deal while we were away and it was nice to take her out again. She walked really well on a lead in the street for a puppy only four months old who hadn't yet started any formal training. Unlike Shandy at the same age she was prone to car sickness but, thankfully, it wasn't too long before she grew out of that little habit.

Sally and Bobby both started training at Beckenham that November. They greeted one another with much affection before getting down to the job in hand. Sally had practised for some two months at home and got the hang of things really quickly. She'd only spent a week in class one before being promoted to class two - no mean achievement for a young lady! We were fast approaching Christmas and the girls were terribly excited by the feverish preparations. Sally was now seven months old and keen to inspect the Christmas tree, with its gleaming tinsel and bright baubles. And Shandy, adult though she was, took great delight in helping her.

* * * *

Three days after Christmas, John Brown, a photographer we'd known for some time, called in to take some pictures of mother and daughter and what better setting, we all thought, than the lounge with a real log fire, Christmas tree and cards all around. Little Sally, sitting with Shandy, made a delightful picture. Some shots were taken of the dogs on their own and others around my feet, where they obeyed the command to 'stay', helped along, I must confess, with just a few biscuits!

The pictures show the great contrast in colour at that time. Shandy was a rich gold, while Sally was pale cream. All retriever puppies go a shade darker within a couple of years, and a look at the ears gives, as a rule, a good indication of the colour they'll become.

The trend today is for pale cream 'goldens', many of whom are very attractive and take the eye of the judges. They also keep a younger look for far longer. But I still have a special weakness for what I think of as an old-fashioned 'gold', with a rich, dark coat. But I would always put colour in third place in a list of priorities when choosing a dog, behind temperament and soundness.

By the end of January 1968, Sally had been promoted to class three at dog school and seemed to enjoy every moment of her training. Shandy enjoyed it too. Her daughter may have been faster to move, but she always worked well at a slower pace. None of us were getting any younger and, that same month, poor Alfred, walking the girls along Crystal Palace Parade, fell on the icy pavement, and broke his ankle, but even so he managed to walk them the long, painful mile home.

* * * *

I took Shandy to an open show in Bexley Heath in March '68, scene of that first fiasco three years earlier, this time under the eye of a different judge. We were both more relaxed than when we had made our debut and I was thrilled when she scooped a third and a Very Highly Commended.

At Easter we sold our lovely boat and I soon missed all the fun she'd given us, at home and abroad. But with a second large garden on the coast and our two dogs, we knew we wouldn't have time to pursue our sailing careers as well. Shortly after this in April, Sally had her first season at 10 months, just two months older than Shandy when she had her first heat. But all went according to plan, with no 'morning after' panics with male admirers.

I took Sally to an open show in Canterbury a week before her birthday in May and, to my great surprise, she achieved a second in our breed class, while Shandy took two third prizes: we were extremely pleased! When the judging was over, I spoke to the lady judge about Shandy and Sally, thanking her for their placings. She replied that she would have liked to place them higher, but 'they are carrying more weight than I would like.' She advised me to cut down the biscuits and step-up the exercise, saying, 'They are a breed which loves to lay about.' I smiled to myself, as the lady in question would have done well to follow her own good advice, being herself somewhat broad in the beam!

The following month, we held a garden fête in aid of the PDSA on a beautiful summer's day in our large Kentish garden facing the sea. The Norwood Guild, of which I had become the organiser, all came down in cars and vans, and many friends helped to run the stalls and side-shows. The local sea cadets' band played on the lower lawn during the afternoon, followed by a children's fancy dress parade. The locals supported us very well, especially considering how new we were to the area and that we were only there at weekends, and only one thing spoiled an otherwise perfect day. Kindly but misguided folk from the guild were seen feeding cake and goodness knows what else to Shandy and Sally, who'd never been known to refuse anything edible. The results were evident later when they were both very sick.

It is always a great mistake to offer anything to a dog without first asking permission from the owner. Kind intentions will not be appreciated in case of illness or special dietary needs. Thankfully, in our case, no lasting harm was done and the fête itself netted a handsome profit of £114. In those days, when most things were sold at fund-raising events for 3d and 6d old money, to take over £100 was a real achievement. The following day, the heavens opened and it was as grey as November. Luckily, Alfred and I had cleared up completely the previous evening but we drove back to London in a downpour, thankful we'd chosen the right day for our big event. But we'd had a close shave weatherwise and we never chanced it again!

A week later Sally was promoted to Shandy's class at the dog club and, as Shandy was in season, she worked every Tuesday for three weeks. It was during July '68,

Sally

almost a year to the day since her mother's success, that Sally followed in her mother's footsteps and won first prize in the open class of the PDSA Exemption Show and second in the Dog of Your Choice class, by public ballot.

I always enjoyed this show, as it was close to home, with a friendly, relaxed atmosphere. We went to several shows that summer and were seldom out of the placings. Looking back, I realise this was a great achievement, in view of our lack of experience in the ways of the show ring, correct grooming and all the rest that ring craft entails. Shandy and Sally were of course pets first and foremost and I often wondered what showing heights they might have achieved if I'd really known my stuff. But even if they'd made it to Best in Show at Crufts, they could never have been loved any more.

* * * *

In our early days with the Beckenham Dog Training Club, our annual competitions were held in November and Sally entered for the first time in 1968. We went into the beginners' class and she worked extremely well. It was I who made the foolish mistake which cost her first prize. When we came to the 'sit and stay' exercise, Sally stayed for one minute, as I knew she would, and even longer, when commanded. When the steward called 'return to your dogs', I, for some reason I shall never understand, gave her an extra command to 'stay sit', even though Sally had made no attempt to move. This extra command drops marks in competitions and as a result we went down to second place. The next class we entered was 'novice' and it was more advanced. The exercises were much the same but the 'stay, sit and down' exercise is longer, a dumb-bell must be retrieved and a round of heel work is completed off the lead. This class we won! Two weeks later we again entered the more advanced club competitions, this time working in Test A. This is done without a lead and with no other command except 'heel' when handler and dog move off. In fact, it is all done by scent. The dog has to retrieve any article belonging to the handler from among many others arranged on the floor without the handler's scent on them. Here again, Sally was working well and, by the time we reached the scent exercise, she was lying in second. However, caught up in the excitement of the occasion, she brought the wrong article back to me and had to settle for third place.

At our annual dance we were awarded the large Chesterfield Cup for our win in the novice class, a medal and diploma for our second in the beginners and another diploma for our third place in Test A. We were very proud of Sally that night and, just for good measure, I even managed to win a box of chocolates in the raffle! The excitement over, we went back to our normal routine, with the girls running in the park every day or walking the length of Crystal Palace Parade with Alfred and me when the park was closed.

Chapter 4
DREAMS OF A WHITE CHRISTMAS

It was the day before Christmas Eve when I drove to East Croydon station to collect my aunt off the train from Eastbourne as she was spending Christmas with us. This gave us a marvellous excuse to drive to South Godstone to have tea at the same hotel and see again those lovely golden retrievers who began it all. Christmas turned out to be a white one, much to the girls' delight, and I was thrilled, too, to watch them running and rolling in the crisp, white snow. Life continued in its happy routine for 'S&S', as I occasionally referred to them in my diary. It is strange how history repeats itself. Sixteen years later I find myself saying this as I make similar notes of outings with our present dogs, Simon and Sarah.

We had, by now, decided to let Sally have a litter and our search began for a good dog who'd already proved himself, not too closely related and, hopefully, not too far away from home. It is always advisable when seeking out a good stud dog to book him well in advance, as well-known ones are quickly snapped up and top breeders don't allow their dogs to be over-used. We called on Mrs Joan Tudor in Horley, famous in the dog world for her Camrose Kennel. We saw two dogs, both at stud: Camrose Cabus Christopher and Camrose Tallyrand of Anbria, both of whom I greatly admired. As with most large breeds, the dog is much bigger than the bitch and so handsome, with his lovely big head and powerful movement. After much debate, I chose Tallyrand of Anbria, then aged nine. Mrs Tudor suggested that maybe Cabus Christopher's puppies were more headstrong and wilful than those sired by Tallyrand. But I now dare to suggest to this experienced and knowledgeable lady that she may have got this the wrong way round! But Tallyrand was duly booked and we now had to wait for Sally's next heat towards the end of the year.

The summer was a very warm one and Shandy was troubled with eczema for the first time. It is quite a common complaint, wet eczema being the worst, and seems to come back every summer. There are a number of helpful preparations on the market for this troublesome ailment, but I found the fastest way to rid the girls of it was to visit the vet for an injection to stop the itching, enabling them to leave the patch alone so that the drying up process can begin, with the added help of a dab of lotion.

One Saturday in August, the RSPCA held a dog show in the grounds of Beckenham Hospital and we decided to go and support it. There were classes for duty and obedience and, when we arrived, we met Bobby and his owner Clare and several other 'goldens'.

'S&S' did not have much luck that day, as things turned out, but Bobby won a class with a fairly large entry. Sally at least kept our end up by taking third prize for obedience and that November we were very proud when she was awarded the Golden

Dreams of a White Christmas

Retriever Cup for the best animal working in Beckenham Dog Training Club during the year.

* * * *

Two weeks before Christmas, Sally came into season and, on Boxing Day, we set off for Horley for the mating with Tallyrand. This time all went according to plan. Mrs Tudor introduced the pair, they had a romp around and, Tally being well experienced in this department, they soon set to work. The couple 'tied' for 25 minutes, which is supposed to be a good sign. and Sally was encouraged to rest for the next 24 hours. A nine-week wait for pups seems an eternity when it begins but it goes in a flash. The pattern for Sally was the same as it had been for Shandy. The first four weeks were treated as normal and then began a slow build up of extra rations. During the whole of that time, there was nothing to suggest that Sally was, in fact, pregnant. She was as slim as ever and, even though she never left any food given to her, she never appeared to be extra hungry.

Early in February, I took her to the vet for some minor ailment and just happened to mention the expected pups. The vet shot a glance full of amazement at me.

'Puppies! What makes you certain that she's pregnant?' he asked. He told me that he couldn't be sure she was in pup as there was nothing to suggest it and he could also hear nothing during a detailed examination.

'What makes you so sure she's pregnant?' he asked again, and I told him I just knew. 'Well, don't put your shirt on it,' he said, with a shake of his head. As we walked back to the car park, I remember saying to Sally herself: 'You are pregnant, I know you are.' And she looked up and seemed to grunt in agreement.

On Thursday 25 February, 'S&S' had a good run in the Crystal Palace grounds and, although everything seemed normal, this proved to be Sally's last run for some weeks. The following day she was restless, so one of the senior partners at the vet's called in to examine her and confirmed that she was getting near her time.

We took her upstairs that night, in case she became distressed prior to the births but the night passed quietly. She refused her breakfast the following morning, as well as her midday meal, and that's when we knew for certain that the pups were on their way. The whelping box was back in its place of honour in the breakfast room and everything was ready. At around lunch-time that Saturday, Sally started to pant and prowl up and down. I was alone in the house, as Douglas had gone to the office and Alfred was at the cinema. An hour or so later, the first puppy arrived and Sally, who had refused to get into the bed provided, had dropped it on the floor, surprised by its arrival. She did, however, break its 'sac', but looked at me as if to say: 'What on earth do I do now, Mum?' I picked up the pup, rubbed it well with a warm towel and held it up to Sally to introduce them, but she still seemed unable to grasp the situation. I was getting worried and, after much coaxing, managed to persuade her to climb into the whelping box and put the pup close to her to be washed. There was a lamp hanging above the bed to keep the pups warm when mum was not around and to keep them even warmer when she was. This time we were taking no chances, remembering the loss of that little pup from our first litter. After half an hour, the second pup arrived and,

A Golden Love Story

suddenly, the whole thing seemed to fall into place for Sally and she settled to work in real earnest. Once again, I dried the new arrival with a warm towel before giving him back to Sally to look after. Then I phoned Douglas to report that the whelping had begun. He decided to take a few rare hours off and came home to help, watching in wonderment as the rest of the litter arrived. A third pup had emerged by the time he reached home and, as Sally rested from each delivery, we gave her a drink of warm milk with a drop of brandy which was never refused. It was around 7.30 pm when the last pup arrived and there were seven in all, four boys and three girls. Their colour range, from deep gold to cream, was all I could have desired and all were around the same weight, except for one girl who was quite tiny. The litter seemed to come in pairs - a boy and girl of around the same weight in each colour, from deep gold to cream, finishing with a little girl in a very pale shade of cream.

Sally knew half an hour later that the hard work was all over now her pups had all been washed and fed, and she settled down to have a well-earned sleep. What a wonderful sight it always is to see an animal mother, contented and fast asleep with her newly born family snuggled close to her. Later on I managed to coax Sally outside to do all that was necessary and she made haste and flew back inside to the whelping bed where she counted each pup, washed them again and settled down to sleep.

This time we were taking no chances and I moved a comfortable chair into the breakfast room so that Alfred could stay with Sally and her family for the night. One never really sleeps under these conditions, but fitful dozing is possible with one eye and one ear constantly on the alert.

I'm happy to say that all was well and quiet after the first night. The vet called the following day, Sunday 28 February, to check on mother and babies and was happy to report they were doing fine. We all had a glass of champagne to mark a double celebration. Alfred was 58 on this very day, having been born in a leap year and usually having to celebrate his birthday on 1 March. The second toast, of course, was to Sally and family.

That evening, we left Alfred in charge and went down to the church at the bottom of the hill. I slept downstairs for the rest of the week, as I was the only one who didn't have to get up early to go to work and could catch up with a snooze in the afternoon. Douglas was more than willing to take his turn as was Alfred but, as far as I was concerned, their offers were out of the question.

Everything continued well until the following Tuesday, when Sally suddenly became restless and distressed. She would attend to the pups and then leave them to wander around the ground floor of the house while she just panted and cried. We sent for the vet and he left sedatives for her. I had a cold myself and as if that and Sally's misery were not enough, Violet was also off sick with a bad chill. The weather was very cold and just for good measure, it started snowing hard! On the Wednesday, we were up all night with Sally because the sedatives were having no effect. I gave her plenty of warm milk, together with several meals, to keep her milk supply flowing. It was my greatest dread that this would dry up but, happily, my fears proved groundless.

Douglas took a morning off to help me: my cold was now so heavy that I would have been in bed had the puppies not kept me on the go. We rang the vet again to report that Sally was still very distraught and had had a bad day. He came over and gave her an injection, explaining that just like a human mother, she was suffering from post natal depression. That night and the following day, things were much better. In fact, our Sally had improved enormously and was almost back to normal, for which we were very grateful.

The pups were now one week old and feeding with gusto. My days were taken up almost entirely looking after Sally and her family. We were now into March and many evenings during the month were filled with various meetings and other commitments.

One evening I presided over a PDSA guild meeting in Norwood and then had to dash away to attend the Beckenham Training Club AGM, hoping it wouldn't take too long and that I'd be home to give Sally her bedtime milk before she settled down with her babies for the night, a job I always liked to do myself. We seemed to be absolutely knee-deep in pups round about this time, for it was during our puppies' second week that I went to see 11 'golden' pups bred in Beckenham by a dog club friend, Peggy Steven, breeder of our present retriever, Sarah. It was, as always, a large and lovely litter, and I went home thinking I'd been let off fairly lightly with only seven.

In March, I went for the fifth year to the Ideal Home Exhibition in March, working on the PDSA stand. Our shifts were in three sections: morning, afternoon and evening, and I always did the Wednesday morning during the show. Fund raising for the PDSA was often enjoyable and in those days the stand was very large, with two ponies on and two off duty throughout the day. They were a great attraction to visitors, needless to say, and we had no trouble filling our collection tins. On arriving home at tea-time, I found Sally reluctant to feed her pups. Violet had phoned our stand at Earls Court earlier in the day to say that Sally seemed disturbed when the puppies gathered round to feed but was calm again once they were asleep. I hoped this didn't mean another bout of depression for her and always, at the back of my mind, feared eclampsia (milk fever), which causes a sudden drop in blood calcium associated with whelping and lactation. However, a quick examination of Sally immediately revealed the cause, putting my mind at rest completely: Sally's teats were very sore from the almost constant sucking and pushing against her tummy by the pups with their little feet in order to get the milk out. Their nails had also grown longer than we realised and must have been very uncomfortable for Sally!

We hadn't come up against this problem with our previous litter, probably for two reasons: Shandy's pups were born in May and they were able to run outside on the concrete in a playpen after a couple of weeks which would have worn their nails down and there were, of course, only three to feed, instead of seven. We quickly covered the affected parts with lanolin and sent for the vet, who gave Sally an injection to ease the soreness from within. We started to teach the pups to lap from a dish but it is never an easy task making the changeover from the ultra-rich mother's milk to something less so. We were introduced to a product called Laughing Dog, which is as near to the natural milk as possible when made up. One by one we introduced the pups to a small dish of warmed Laughing Dog and they soon began to lap, much to our and Sally's great relief. Before putting the pups back to Sally, we gave them a good drink and cut their nails and there was no more trouble for the rest of the natural feeding time.

A Golden Love Story

Several visitors called that week to see the new family, among them Clare and John, who'd had Bobby, of course, and now wanted one of Sally's pups. Nothing had been decided at this stage - they simply wanted to see the new babies and admire. Two more visitors arrived who'd been disappointed not to get a puppy from our first litter: George Madden, who managed a fish shop in Sydenham called Tutts, and his wife Nora. We were among George's regular customers and almost always had Shandy with us in the car when we went shopping. He was very taken with her and he and Nora asked if they could book a puppy. If Shandy hadn't lost her little son, he would doubtless have gone to the Maddens. So, when Sally was mated, they went straight to the top of the list, just in case.

Many dog club friends, themselves breeders of 'goldens' came to visit. Breeders, for obvious reasons, tend to be a critical breed themselves, so I was particularly pleased when one plain-speaking lady took a look at our seven bundles of fur - now very active, alert and nicely plump - and exclaimed: 'What a lovely litter!' She went on to congratulate me on rearing them so well at that stage.

The pups were weaned at a month old. Their four meals a day from me consisted of a very finely minced meat at one meal, with small biscuit meal soaked in vegetable water, fish for another and a milky Farex meal for another. But Sally still gave them a night time drink of her milk and continued to do so for the next two weeks. After that, I increased the solid food to five meals a day.

Soon we decided it was time the pups were photographed for the first time. They were getting to the stage when they looked at their most appealing and John Brown agreed to come and do the honours. One thing of which we were certain was that there was no way he would get those seven bundles of fur to keep still on the floor, or even in the playpen long enough for him to get a decent study of each one in a group shot. But it was during this session that I discovered photographers have a number of tricks up their sleeves for such occasions and will look around for nooks, crannies and utensils of every kind. After taking in the bedlam in the playpen for about 15 minutes and looking round for some way to cure it, he spied fitted cupboards in the breakfast room on one wall, part of which was a pull-out work top. We pulled this down, laid a blanket on it and sat each pup down with a dish of milk in front of it. Perfect! John kept clicking away all the time while the pups were lapping and, as they looked up, he managed to get yet another great shot.

On 12 April, the Simmonds returned to pick a puppy - a difficult job, as by now all the pups were much the same size and colour, except for one little boy who was a rich, dark gold while his siblings were mid to pale gold. After much discussion they decided on one of the paler boys. His name was to be Salvador. The Maddens had decided on the mid-gold boy, a little larger than the other three, and they called him Jumbo. We avoided giving any pet name to unsold pups to prevent confusion later on and had no idea just how soon the litter would be completely sold. It had only ever been mentioned to friends and dog club connections and not advertised in any newspaper because that way we hoped we'd find well-recommended owners. It's also always helpful to a new owner when the pups are named before they leave for their new homes as they can learn their new names in readiness. I realise, of course, that this is not workable for a big breeder with a large kennel whose puppies are reared outside. What is more, I never cease to wonder how they know which buyers have chosen whom from a large litter or maybe more than one, as it is quite common for a big breeder to have several puppies from different mothers at one and the same time.

It was mid-April when a friend in Shirley phoned to say some neighbours were looking for a golden retriever puppy. The daughter of the family was getting married and mum and dad felt they would like the companionship of a young puppy to help fill the space in their lives when their daughter left home. Mr and Mrs Wagstaff came to inspect the litter and chose a boy. They were a nice, middle-aged couple, Mr Wagstaff having just retired, and I had no doubts about them as new parents for one of our beloved babies. They chose a light gold boy and were going to call him Lovell after one of the astronauts who'd that morning safely splashed down in the spaceship Apollo after the first landing on the moon.

But he was only called Lovell for a week or so before Mrs Wagstaff phoned to say she'd found an easier and more attractive name for her pup. She preferred Koppa, which she thought might be Russian for 'gold'. I agreed that it would be very appropriate and much easier to call out in the park, so Koppa it was. It was a good thing our visitors arrived when they did. Only the day before the vet had come to worm the pups for the first time and, two days later, one or two became upset by the medicine and in no mood to greet company, buyers or not. With all the extra work this entailed I was soon likewise in no mood.

But, having got them over this little trauma, the vet came back a few days later to give all the pups an injection against distemper. This was because this litter, unlike the first one, wouldn't be leaving at eight weeks. Summer was drawing on and most of the new owners had holidays booked and were loathe to take their puppies for a few weeks, only to leave them with friends or in kennels so soon after adapting to their new homes. So they stayed with us, some for far too long, I felt, but it seemed to make little difference when they did eventually move on.

* * * *

One morning around this time, a Wednesday to be precise, I went to the hairdressers and on to do some shopping later. During my absence, Violet made a frantic telephone call to Douglas in his office telling him she could not find Samantha anywhere in the house. Douglas tried to calm her as best he could, asking questions about open doors or gates etc. Violet assured him all doors leading out of the house had not been opened since my own departure that morning. Douglas then suggested she make another careful search, he was extremely busy and did not want to return home early unless it was absolutely necessary. About half an hour later, Violet called to say all was well; she had found Samantha curled up fast asleep in a little corner on the landing; being dark there, Violet had failed to spot her the first time. She, (Violet) then sat down with a cup of tea and an aspirin to try and rid herself of the headache the incident had given her.

Chapter 5
SPRING BY THE SEA

It was the end of April and Douglas took one of his rare days off, relieving me of my puppy-minding duties and leaving me free to go the Tamarisk for the day to do some spring cleaning with Violet. Tony Nicholls, Douglas's driver, was able to take us and give me a really welcome break.

Meantime, Douglas fed the pups with their morning meal of porridge with honey, giving them quite a large helping. He then settled them down, or so he thought, took a leisurely bath, pottered about, read the post and finally made his way back to the kitchen for a cup of coffee. To his horror, Douglas found the fridge door wide open and saw that a 7lb chunk of beef had been pulled out by the pups and devoured. There they were, lying all over the floor, fast asleep, with tummies like footballs. Douglas decided to phone the vet.

'What shall I do?' he asked, after explaining his plight.

'Congratulate them,' said the vet, 'and don't feed them for 24 hours.'

'Phew,' breathed Douglas, ' that's a relief. They've eaten every scrap of meat in the house.' The vet reassured him that the pups would sleep, spend pennies, etc., and sleep again, but they wouldn't die. A small puppy can eat a vast amount for his size with no danger. He was right, of course, and no harm was done. As the dogs all lived to a good age, perhaps the feast even stood them in good stead.

* * * *

We were now into the first week of May and the weather was really warm. For the first time that year, I sat out in the garden in a sunsuit, with the playpen on the lawn, and watched our little family at play: a scene far more familiar in high summer. During that week we discovered that one of the girl pups had dandruff - the tiny girl in the litter. The rest of the litter seemed unaffected whenever anything went wrong; it was almost always little Shalima who suffered, (yes, this was going to be her kennel name at the time, though later changed to Starlight Shannon). Our vet gave us some special shampoo with which to bathe her and the trouble soon cleared up.

One Saturday in May, we'd decided to leave for our seaside home for a week's holiday. There was never any question of us going anywhere without the entire canine family and so we set off with grandma, Shandy, sitting in the front, of course, and Sally, her seven pups and me in the back. Most of the pups settled themselves on the ledge at the back so they could peer out of the window and it must have been quite a sight for motorists following us to see five or six small puppies of various colours bobbing up and down in front of them. There was always a reaction from cars that overtook us, from smiles of delight to expressions of sheer disbelief. The journey, on the whole, was really quite good, all things considered. I'd embarked well prepared, with plenty of old towels

Spring by the Sea

and newspapers and, sure enough, all the tiny tots were sick a couple of times, while the tiniest of all (our dear little girl) managed it no fewer than seven times. This upset me very much and I did my best to comfort her on my lap throughout the journey. Luckily, she slept quite a bit and I was very thankful when we finally arrived.

As we drew up to the house, I reminded Douglas that under no circumstances were the pups to be put down on the ground until safely inside the garden, as their injections hadn't yet been completed and any dog could have been spreading germs on the roadside. We got round this problem by standing Douglas inside the gate while I stayed inside the car and handed the pups to him through the window, one by one. He, in turn, made the trip from gate to conservatory, depositing each pup inside and quickly closing the door until the seventh one was safe and sound. What a performance!

Once they were all together in the conservatory - which had been prepared in advance by removing chairs, tables, plants and floor covering, the puppies let rip. But they quickly flagged with the excitement of it all and fell asleep on the tiled floor, warmed by the sun on the windows and glass roof.

As it was May and not yet too warm, our babies were in no danger of being overheated but we opened the windows to allow in plenty of fresh air as they slept. We wouldn't dream of repeating the exercise in high summer, as such places quickly heat up to intolerable temperatures, as does the car. I always dread the thought of summer when I think of how many thoughtless and irresponsible people leave their dogs in cars with or without the windows open. They may start off by parking in a shady spot but don't they realise that the sun moves around? They should try sitting in the car in the same conditions and imagine they can't open the window or call for help. On more than one occasion, I've left a note on a car containing a dog I considered to be in danger in high temperatures and would urge any animal lover to do the same thing. You may run the risk of being branded a busy-body, but what does that matter if a dog is spared a bit of suffering in such conditions.

The little family behaved beautifully in strange surroundings for the first two days and slept at night in a big box in the kitchen. The first full day at Tamarisk saw them out on the lawns, chasing each other with great glee. This was their first experience of a big, wide open green space and they were intent on making the most of it. But, as there are two ponds, deep enough to spell trouble for tiny tots, playtime had to be constantly supervised.

The second day turned very chilly, with a cold wind blowing directly off the sea. Our home faces due north, so when the wind blows off the sea, it is very cold indeed. They continued to sleep in the kitchen but we decided the conservatory was the best place for the pups during the day, until the weather improved. This decision was probably to blame for the situation Douglas discovered when he went into the kitchen the next morning.

The pups had plenty of room to play inside, but there is nothing quite as tiring as a good chase around a big open space and, good as they were that night, the next morning they were awake very early and creating mischief.

A sight for sore eyes awaited Douglas behind the kitchen door. All was suspiciously quiet but there, in front of him, was the most dreadful mess he'd ever seen. The family, up and about first thing and looking for something to do to occupy their time until breakfast, had found a sliding cupboard door under the sink, open by about two inches. Inside was a large heavy bag of grey cement mixture and we can only assume that two or three of them got together and pulled it out. The entire room was covered in the stuff, and that included the pups themselves. They were having a splendid time paddling in it and kicking it all over the room and each other. I leapt out of bed at Douglas's anguished cries and, without stopping to put my dressing gown and slippers on, hurried to the kitchen.

My first worry after surveying the disaster area was whether any of the puppies had swallowed the dust. I picked each one up to examine the mouth, wipe it out with a clean cloth and hope for the best. Then we put them all in the conservatory while we set about cleaning the kitchen.

We finished around nine, long after the pups' breakfast time, as they noisily reminded us from the conservatory. We gave them a normal breakfast of porridge, milk and honey and tried to be optimistic. Would they, perhaps, all turn into statues before lunch-time, a bit like Lot's wife becoming a pillar of salt? I was quite worried that some of the mixture might have found its way into their little tummies, which could have disastrous results. Should we call the vet now or wait? Vets are busy people and, while we always go to them in a real emergency, we hesitated to call them out on what might turn out to be a fool's errand. So we decided to keep a watchful eye on things and hope!

That day, four friends were due to visit us, all dog lovers who understood the situation. We'd decided in advance to take them out to lunch in a nearby restaurant. Trying to cook a meal for six in a small kitchen with seven lively pups, plus mum and grandmother around, was really not on. We gave the pups their normal midday meal, let them have a good run on the lawn near the house, settled them down again in the kitchen and this time securely closed the cupboards, tying some of the doors up for safety's sake. Then we set off for lunch. On our return, all appeared well and when the rest of the day, as well as the following one, passed without incident we finally began to relax.

* * * *

The next few days went by peacefully, the weather varying from warm and sunny to cool with downpours, and the pups exercised accordingly.

Douglas had to go to a board meeting on Friday, so I had to cope with the frustrated family cooped up during the rain which lasted all morning. Our week's holiday over, we drove back to Sydenham and all the pups were again sick on the journey home. The weather was now suitable to put the pups out in the garden for the afternoon, at least, and the combination of sun, air and play made for long quiet nights for which we and Sally were truly grateful.

Douglas's aunt from Herne Bay had recently been widowed and she rang that Wednesday to say she'd like a puppy. We were very pleased and arranged for her to come and inspect the litter the next day. John arrived too, to take the second batch of photographs, this time in the garden with, hopefully, mum and grandma.

I won't give a blow-by-blow account of the chaos that ensued. John had decided to put a plank of wood on a rustic garden seat with a nice leafy background so he, his

Spring by the Sea

assistant and I spent a full half-hour in rounding up the litter, sitting Shandy at one end of the plank and Sally at the other.

We finally got all the family into position and, using every ounce of his expertise to hold their attention, John set the camera rolling and kept it going until the pups simply got bored and jumped down. It was all worth it and we got some excellent shots with interesting expressions on the pups' faces, matched by the bewilderment on the faces of the two doggy adults in the party.

It was quite difficult to make a choice when the proofs arrived, but we decided on one, which was duly enlarged and hangs in the breakfast room, the birthplace of all our puppies. Even now, it is high on the list of my most treasured possessions.

Auntie Betty arrived with Douglas during the evening. She decided, after much discussion, to have the 'mid-gold' of the litter and to call her Midge. I felt relieved the puppy was going to a member of Douglas's family who'd owned and bred dogs and also that I would be near enough, when in Kent, to keep in touch and follow her progress.

But our pleasure was short-lived when, on the following Tuesday, a letter arrived to say that, after much thought, Betty had decided not to have the puppy in case she couldn't cope if she ever fell ill. As she lived alone, she felt that it would be unwise to take on the responsibility. No doubt this was a sensible, if sad, decision as a large energetic dog can be very exhausting for an elderly person.

* * * *

Strangely, that very evening, John phoned to say that he, his wife and family would very much like to have one of the litter - 'a girl dog, if possible,' was the message. They came back to Elmwood and, without knowing anything about Auntie Betty, chose little 'Midge'. We were delighted, knowing she would be going to a lovely home. This time she was to be called Salvia, partly from her kennel name and also because of her mother. Mr and Mrs Wagstaff called in the same day to see how Koppa was getting on and, the following week, Mr and Mrs Madden came to see Jumbo, who had been aptly named as he was, even at that stage, the largest boy in the family. The following Thursday, I received a call from Milly to say that Mandy, Sally's little sister, who'd been mated to the same dog, was due to whelp. It was the first time Milly had bred from her dog and I offered to go to Hayes to lend some moral support. Mandy was in no hurry to produce and I stayed on into the evening, long enough the see the first puppy born, arriving back home at nine. I was just in time to help Douglas entertain four friends for drinks, all dog lovers and I quickly discovered that John and Ethel Gregory, both there that evening, wanted to have our dark gold boy Rusty. He was a litter twin to one of the girls we had decided to keep, our Samantha. They both had the same colouring and expression, although brother Rusty may have had a nicer temperament when it came to other dogs.

Milly phoned late that night to say that Mandy had had three boys and three girls. It had been a difficult whelping, but all was now well. What a day!

A Golden Love Story

Clare rang to say they'd love to collect Salvador that Sunday and, although I agreed, I felt the first pang of heartache at being parted from my babies. I missed each puppy greatly on his or her departure for quite some time, and it's a very odd feeling even when you are keeping one.

When the last puppy has left, it feels just like those ten little Indians in the rhyme - and then there were none - and I always found it very difficult to cope. It is, of course, somewhat different in a large kennel, where the relationship is more remote.

Sunday came all too quickly and, as before, Clare and her family arrived with vet, David Simmonds, to inspect the new member of their family.

Chapter 6
GOODBYE, SALVADOR!

Salvador was the first to leave home. He went at tea-time to join Bobby, his uncle, in Beckenham. I phoned later to find out how the meeting had gone. Uncle Bobby, it seems, had greeted his young nephew with great delight and was most protective towards him. The following morning, Clare telephoned to tell me that the pair had settled in well overnight and, as the weather was quite hot, they'd played outside early, in the morning cool, blissfully snoozing afterwards in the lounge.

The following day I called to see Mandy and her mother and sister. Then, two day's later, on Thursday 11 June, John and Sheila arrived in the afternoon to take little Salvia home. Her kennel name was Sunspray Salvia Silver Star and she was to be their bright, shining star for more than thirteen years.

I was glad of the extra company when Ethel Gregory and the children dropped by that evening to see little Rusty. But all through the chatter and the laughter, I couldn't help wondering how our tiny Salvia was getting on in her new home. The boys were really beginning to look like young dogs now but, as far as I was concerned, they were still our babies. The pups were now four months old, all injections completed, and we could venture into the park for the first time with the four remaining in our care.

This first run in the park is always a heart-stopping moment for me. At 8 o'clock on Sunday morning, we set off with Shandy, Sally and the little ones. They were quite good on the lead and we made sure, once again, on the short walk from house to park gate, that we didn't let them off until we were well into the park and away from all outlets to the main road.

Once there, we slipped the leads and held our breath. They fled in all directions, intent on exploring every inch of this new exciting place. We did note, however, that they didn't stray too far from mum, grandma or us, and all came back when called, at least the first time. On the second call, everyone came bounding back except Koppa, who obviously intended to explore more fully, and solo to boot! He probably thought that if he came across a good find, he could keep it all to himself. I had quite a chase on my hands until he finally gave in and, on arriving home, there was little doubt who was more exhausted as I collapsed through the door. On the following Tuesday, Mr and Mrs Wagstaff arrived during the afternoon to take him home. Mr Wagstaff had just retired and liked walking so Koppa was due to receive all the attention of a newborn baby in that house. I remember handing Mrs Wagstaff Koppa's diet sheet and her remarking jokingly to her husband that, when fully grown, the dog would be eating around one and a half pounds of meat a day. 'Very well, dear,' he remarked drily. 'It will be cheese for you and meat for Koppa.' And they set off for home around tea-time.

A Golden Love Story

I missed my baby terribly that evening and he was my first thought on waking next day. I waited a whole hour before making the ritual call to enquire how the first night had gone. 'Splendidly!' came the reply. 'He's so good,' cooed an ecstatic Mrs Wagstaff. My mind at rest, I prayed their contentment would continue, and it did.

Mr and Mrs Madden came to take Jumbo home on Saturday. He would be living the nearest to us and I knew we'd be seeing a lot of him when we went to the fish shop in Sydenham about twice a week. Even so, the evening we parted was quite a wrench. But the Maddens had been dog owners in the past, and were true dog lovers. As with Koppa, Jumbo was to be the baby of the house, very much loved, but, in his case, finally indulged to a sad and unforeseen conclusion.

The following day, John and Ethel came before lunch to collect Rusty, the lovely rich, dark golden boy who was now a real twin to our own Samantha. We knew we'd be kept informed of his progress and saw him frequently over the years. What very handsome boys they all were. And I know, to my great satisfaction, the pleasure and loyal companionship they gave and received. The same goes for little Salvia, the only girl in the litter to leave home.

So now it was just Samantha and Shelley (who started life as Shalima) at home with us, together with Shandy and Sally and they settled quickly into a routine again. We had planned to keep only one puppy and, somehow, I felt guilty about having two. Perhaps we were depriving someone of a pet, and yet I felt Shelley was so small and vulnerable she needed our care. But fate was to prove me wrong and I was to wish very much that she'd gone safely to a caring home. Three telephone calls came on the same day: Jumbo, Salvador and Rusty were all doing well and had settled down quickly. Back at home, the two girls were getting along nicely and were joining us in the lounge during the evening and, no doubt, feeling very grown up. Mother and grandmother were on hand to give a lead and little Samantha and Shelley soon found their way around the layout of the house, and us! On Wednesday evening that week, Sally returned to the dog training club for the first time in five months. Samantha came with us for the first time and, like her mum, seemed quite at home watching Sally and all the other dogs doing the exercises. History was repeating itself once more.

Everyone was delighted to see her and a great fuss was made of her. The next Tuesday was very hot, so we ran all the dogs early and settled them in the cool of the house for the rest of the day. That afternoon was one of the nicest for a long time when I called on Milly with a friend to see how Mandy's puppies were progressing.

They were over a month old and looking fine, a great joy to both Milly and Mandy. And if all that pleasure wasn't enough, I took Alfred that same evening to meet some more golden retriever puppies belonging to a lovely mum named Cilla, in Beckenham. They were about the same age as Mandy's little ones and all so very appealing.

* * * *

On the following Monday, Shandy, Sally and I left together for a short break at Tamarisk on our own for two days, leaving the pups at Elmwood. The weather was lovely and we made the most of it, on the beach and in the garden. But the following day it turned foul, pouring with rain non-stop and blowing a gale late into the evening.

Good-bye Salvador!

The girls got a good soaking in the morning and spent most of the day snoozing while I cleaned out the conservatory. Every night since they had been born, both Shandy and Sally had slept undisturbed, come thunderstorms, gales, break-ins, you name it. But, this particular night, they refused to settle and ran around barking and growling almost till dawn. I could only conclude that an intruder must have been prowling around the garden.

The next day we returned to Elmwood and it was good to be back with the family again. The pups had been very good girls in our absence so we took them all for a run out to Felbridge, Blindley Heath and Godstone next day. Little Shelley was very good in the car but Samantha clearly didn't like it and would never really be happy about travelling. We stopped for tea in Godstone at our favourite place where we'd seen our first 'goldens' and where there was a PDSA collection box that I serviced from time to time. These trips were a great source of enjoyment for all concerned - even Samantha, once the car had stopped moving. Saturday was fairly hectic, with a packed programme. First thing in the morning I had to arrange the flowers in the church - something I've done for twenty years on a Sunday between my late mother's birthday and wedding anniversary which is the same month in which she died.

In the afternoon, I took Sally to the PDSA Dog Show in nearby South Norwood. But this was a mistake. We may have supported the charity, but we were also wasting her time. She wasn't yet in good condition, following the birth of her family five months previously, and I soon found out that she wasn't prepared to work in 'obedience' just yet. It was one of the few times we came away from a show or competition without an award of any kind.

Finally, that evening we were invited to a party down in Kent. It was of course unthinkable that we should leave the girls alone in the house so, as Alfred was away on a holiday, a member of my fund-raising guild came over to take care of the dogs and the house.

We left her some supper on the table in the kitchen and, during the course of the evening, the phone rang. Nellie got up to answer it and was away from the table only a few moments - just long enough to find, on her return, that all the meat had disappeared from her plate and only the salad was left.

Eight eyes, full of innocence, watched her reaction as she gazed at her plate in disbelief. There was no indication as to who was the culprit. Nothing like this had ever happened when we left food on the table. So no doubt one of her charges had taken full advantage of the fact that a stranger was looking after them. There was nothing poor Nellie could do as it is always pointless to scold a dog after the event. It must be done, within reason, when caught in the act.

The following week I took all the girls to visit Bobby and Salvador in Beckenham and it proved to be a hectic afternoon. They had a great game in the garden and I was glad to see Salvador was growing up well. That Saturday saw Shandy at a dog show in Tulse Hill, which was run in conjunction with a Conservative fête. Shandy won first prize for Bitch and Best Condition but, when it came to the child handling class, she refused to move off around the ring with her unfortunate child handler, obstinately determined to get back to me at all costs.

A Golden Love Story

Jumbo came to see us that week. He was a really lovely animal, the pride and joy of his owners and, as with his brother, Salvador, and Uncle Bobby, he had a wonderful time with his family in the garden.

Chapter 7
PUPPIES' PROGRESS

It was Wednesday 2 September 1970, when Shelley went to dog training class for the first time at just six months old. She did well - her practice at home had obviously paid off and she took to it like a veteran, as did sister Samantha on her debut a week later. I'd certainly jumped in at the deep end, training two puppies at the same time. It meant, of course, that they had to go on alternate weeks until one was ready to move up a class. We didn't have long to wait. Sam was promoted to the second class on 14 October. Jumbo had also started coming to the club and was doing very well indeed.

That summer had been a good one - even the weather had been fine. On the very hot days we ran the girls as early as possible in the Crystal Palace grounds or on the sea front in Kent. If we had an outing during the day of only two or three hours, we always left the pups to rest in the cool of the house on their own. But, if it was any longer, someone always came to feed and let them out. I am greatly disturbed by the number of hours many dogs are left alone at home and, even worse, by their being taken out for the whole day in a car in hot weather. I fervently wish our national animal welfare societies would give more publicity to these unhappy practices.

We made a few trips to Beckenham Recreation Ground so that the girls could have a good run and romp with Bobby and Salvador. The little family playing together made quite a sight, and there was never a cross bark between them.

* * * *

It was October and Shelley had also graduated to the second class at the dog club. I wanted to try to work both pups each week, so I took one and different handlers took turns in taking the other. This arrangement worked quite well and kept both in training for the basic heel work at least. There was a beautiful Indian summer to enjoy. Our one big disappointment was the postponement of our planned holiday to South Africa, due to take-over talks at Douglas's firm, making it impossible for him to be out of the country at that time. By the end of the month, however, the take-over had gone ahead and he had been made Managing Director, so our holiday could now be put forward to February 1971, even though it meant we'd be away for the pups' first birthday.

The dog club's annual competition took place in November and, on the 11th, the pre-beginners' class turned out to be a real family affair, with Jumbo gaining second place, Shelley third and Samantha fourth. As we stood in line to shake hands with the judge, a round of applause broke out and the family went mad, sensing the congratulations were for them. I was delighted when two yellow and green rosettes were presented to me in recognition of this achievement at our annual dance a few weeks later.

A Golden Love Story

* * * *

I called to see Koppa in Shirley around this time. He was now a lovely big boy, with bags of energy, and Mr Wagstaff told me with a grin: 'I take him every day to the woods for a really good run, hoping to wear the blighter out, but the only one to come back exhausted is me! His energy knows no bounds.'

Christmas was again a white one and we awoke on the big day itself to find it had snowed heavily in the night. The girls loved every minute of it. We were up early to get the turkey under way and I went to church with Douglas and my aunt, whose birthday it was. Afterwards, we took the girls to the park for a wonderful romp in the deep snow. It was always a pleasure to see them run and plough through it, burying their noses and rolling over and over. Samantha was a great little roller but, as the snow was new and amazingly clean, they all arrived home sparkling bright. Friends were due to lunch, and would stay on till after supper, so I was grateful there was no big cleaning-up operation to go through first. We decided to give the pups a second treat that sunny afternoon, so Alfred and I took them back to the park while Douglas and our friends enjoyed a snooze. I wouldn't have missed the puppies' joy at being back in the snow for a second helping for a hundred snoozes.

We repeated the dose the following day. More snow had fallen during the night, so it was even thicker and more fun than ever. Then it was back to a roaring log fire and our turn for a doze after a nice cup of tea. Bathed in the soft glow of the Christmas tree lights, the girls made a lovely sight around the hearth.

The day after Boxing Day saw more snow falling, and quite hard, but Douglas, Aunt Eva, Shandy and I still decided to drive down to Kent. I thought how lovely Tamarisk looked under a blanket of snow which was, by now, very thick indeed. The house is so close to the sea you can throw a stone into it from the garden so it was rather strange to see snow lying so thickly for so long - I would have expected the salt in the air to thaw it quickly.

Shandy was grateful for another chance to run free in the luscious, white stuff again, just as she had in the park. She was never a roller, always appearing rather stately in carriage. That was the sort of antic she left to the younger, less dignified generation. There was always a look of surprise and mild disdain on her face whenever she watched her offspring rolling around on their backs with their legs in the air. We had a happy two-day stay, despite the conditions, with plenty of runs for Shandy along the sea front during daylight hours.

Evenings at Tamarisk saw her curled up, not in front of a big log fire, but in front of a gas model with log effect. But the first time she saw it she seemed to decide that this would do nicely and settled down to sleep as near as she could and we would let her. Left to herself, Shandy would stare into the fire, but this is not good for a dog's eyes and affects the condition of the coat.

When we arrived at Tamarisk, our great friend, Gladys Dadd, came round with her family for a drink. Shandy loved people more than dogs and was never happier when visitors arrived and she could show off her power to retrieve. If it was something which belonged to the visitor, so much the better! Gladys, who had a small poodle called Mimi, was a special friend of Shandy's, and she seemed to regard her as part of the family, what with her car being stowed away in the garage and her offers to take care of the bungalow when we were away. Dogs, in my experience, always have this wonderful way of knowing who does or does not 'belong'.

Our second night saw us at the panto at the Marlowe Theatre, Canterbury - minus Shandy, of course. I didn't tell her that the production was *Puss in Boots* - she might not have been amused. But she always accepted that we'd be away for a while - three to four hours at the most - and, on our return, the greeting would always be the same. Whether it was a night at the theatre or a ten-minute trip round the corner, she always went wild with delight.

We returned to Elmwood two days before New Year's Eve. It was bitterly cold, too cold even for snow, but the house was warm and spotlessly clean. The three girls who'd been left at home with Violet on this short visit were all well and more than happy to see us back safely again. The New Year came in with the weather even worse. Fog had joined forces with the snow and ice but, by the end of the first week of the New Year, it gave way to a rapid thaw, making the streets muddy and slushy. Cleaning the girls after a walk, with their thick coats and feathering, was no mean task. Thank heavens for radiators, plenty of old towels and newspapers spread on the breakfast room floor! They are ideal for preserving the rest of the house from muddy paws and for absorbing all the dampness. During the first ten days of January we noticed that Shandy was having some difficulty in getting up again after resting. The trouble seemed to be in her back legs. We took her to the vet and he had no idea what was causing the trouble, suggesting an x-ray the following week. When we collected her afterwards, together with some tablets, she was still sleepy and we were told that arthritis had been diagnosed in the spine and back legs. I felt this was a bit young for such a disease, as Shandy had only just turned six but, on speaking to other people with dogs, found it was quite common. I still felt very concerned and wondered how long Shandy would have to live with this distressing ailment and how progressive it was likely to be. For the moment, however, the tablets seemed to be doing the trick and she was back to normal, returning to her family at the dog club just two days later.

Koppa, who was by now a fine golden dog with a very soft expression, came down to have a look but Mrs Wagstaff had decided not to pursue any formal training for him, feeling they could manage him to heel for their needs. We left on our three-week trip to South Africa on 13 February after walking all the girls in the park and leaving them in Violet's capable hands. It was a night flight and we left for Heathrow at six. The worst part was saying good-bye to our girls for the next three weeks. The luggage was loaded into the car, giving rise to much excitement, and I quickly hugged and kissed each one before making my exit. It was not too bad after all, once underway. We knew our pets were in good hands in their own home with people they knew and loved and everything would continue for them without interruption.

* * * *

We greatly enjoyed the beauty of South Africa. It had long been an ambition of mine to visit the Kruger National Park and other game reserves and I was not disappointed. Our guide told us we were particularly lucky to see 26 different species on our trip into the park in one day. The park covers an area larger than Holland and many tourists enter the park and don't see one animal all day!

We docked in Cape Town on Sunday 25 February, having come by sea on the SS *Vaal* from Durban. I wished I could have been home that morning partly because I'd been suffering from a bout of sea sickness, but mainly because it was the pups' first birthday. I thought about them quite a lot during the day and tried to picture the routine at home, bearing in mind the time change. On our arrival in Cape Town we were taken to visit the PDSA animal treatment centre and were much impressed by the conditions and standard of work carried out. I was happy to leave some of the Christmas gift tags which I make out of old greetings cards to sell for the organisation's funds each year. Hopefully, the staff were able to sell them and induce others on the fund-raising committee to do the same.

We arrived back in the UK and were collected from Heathrow with all the news. The postal strike, which had begun before we left, was still on and decimal currency had been introduced. I had to cope with that on my first trip to the shops, but was surprised to find how easy it was, even though I felt I was still abroad! The girls gave us a wonderful welcome, but Shandy was limping badly, finding it difficult to jump up at us. Apparently she'd been like this most of the time we'd been away. Another surprise - or rather a shock - awaited me outside the bedroom window. I found myself looking down at a large area of grey concrete which had once been a lovely green lawn. The two oldest girls had started 'spending pennies' on the lawn, causing large brown patches to form, and the two later arrivals had just about finished it off, turning the lawn into a large muddy patch. (Strangely enough, the three large lawns at Tamarisk were never similarly affected).

Douglas had instructed workmen to come in and concrete the lawn over while we were away, but the job was very badly done. A public car park would have looked better, and several refinements had to be made to the edges before we could even begin to live with it. We don't have the same problem with our present two bitches, Sarah and Selina, as their diet is very different from their predecessors', who were fed entirely on meat and biscuit meal, which made their urine much more acidic and so tougher on the grass.

* * * *

There was a gathering of the clan in April at the PDSA dog show in South Norwood for Sally, Samantha, Jumbo, Koppa and Rusty. It was Jumbo who came away with an award that day, when he took a third in obedience. As Koppa and Rusty hadn't had any kind of training for the ring or in obedience, Jumbo's comparative success wasn't surprising. Even in what was then considered a small exemption show, held in aid of charity, competition is keen and dogs need plenty of practice to show themselves at their best. Learning to stand still for a few seconds and allowing strange hands to run over you is the key and this takes time to perfect.

Shandy stayed at home this time with yet more tablets to try to relieve her condition. She was always a great favourite with the local shopkeepers and especially with the lady in the health food shop who always waved as we went by. One day we called in and she said she'd seen Shandy out with Alfred, while we were away, and had noticed how badly she was walking. I asked if there was any natural remedy she thought might help, although I knew that there was no cure.

'Culpepper Arthritis Herbs have helped a great many people, some of them bad cases,' she said. 'Why not give them a try and see what happens?' These herbs are usually taken in the form of tea, but we gave them to Shandy sprinkled in her food. 'Don't expect any results for at least three months,' we'd been told. 'But don't give up.'

Shandy took the herbs for about five months and there was certainly an improvement. She lost her limp and ran upstairs with little effort, was able to jump into the car, which she adored, and on to the beds too, when she thought no-one was looking. She was never again treated by the vet for arthritis and remained on the herbs for the rest of her life. They were sometimes hard to come by and also became more and more expensive, but we tracked them down wherever we could.

Chapter 8
GROWN-UP GOINGS ON

The pups, who were now fully grown, made a striking combination. Age had seemed only to highlight the two contrasting shades - Sam, so richly dark golden, Shelley, a pale cream. They both seemed fit enough, visiting the vet only for minor things like ear cleaning or eczema.

One day in July, our dog training club was giving a demonstration in conjunction with a fête for London Transport. Sally had come into season six days before and couldn't go, so Shelley was chosen to go in her place. Working in the same team as her brother Jumbo, she did very well. Then, a few days later, Samantha was promoted to the third class.

The summer of '71 had been a good one on the whole, with reasonable weather, and the pups and Shandy took it in turns to come in pairs to visit Tamarisk. Sometimes all four joined us if we were staying a full week or more.

Douglas and I spent a day in Eastbourne during November, visiting his mother, who was in a nursing home after an eye operation. The 'ladies' stayed at Elmwood Cottage, where Alfred ran them, fed them and settled them down before going out. We returned at around six when it was already quite dark as the clocks had been put back the previous Sunday. As we waited in the middle of the road to make the right turn into the driveway, I spotted two young men struggling out of our gates with something heavy and realised that our TV, enclosed in a cabinet, was being loaded into the boot of a car. I yelled to Douglas: 'We are being burgled,' and leapt out of the car in a futile attempt to stop the men. My next thought on entering the house was: 'Why is it so quiet? Where are the girls?' I made a dash for the kitchen and breakfast room where, thankfully, our four 'goldens' were stirring from a good sleep after their dinner. So much for our guard dogs!

Despite various seasonal bugs, we managed to keep the dogs' routine going between us, and I managed to recover my good humour in time for the Christmas festivities, only to be laid low again with a stomach upset on New Year's Eve. Alfred had also been ill over Christmas with flu, so he was unable to help with the exercising. Perhaps the New Year will bring a better phase, I thought, as 1972 dawned, and I sat disconsolately in bed. The New Year certainly hadn't begun well!

Our lovely big budgie, Mr Bumble, had died during the night, which I suppose was only right and natural, as he was 19 years old. He was given to us to look after by a good friend who was chief veterinary surgeon to the PDSA. The bird had originally come into his care from one of those famous stage families of yesteryear, the Lupinos. Mr Bumble had, at what was already an advanced age, been given a young lady companion in her prime and she'd tried to get the elderly gentleman to mate with her. Poor Mr Bumble was plagued almost out of his life and the owner had sent for our veterinary friend, Holland Ffoulkes, to seek his advice. This was, of course, separation. Mr Bumble was too old to cope with an amorous and youthful young lady and so he came to us, the only other creature ever to live with us.

When he arrived, the dogs' reaction to him was the same as to anything new to the house. They liked to inspect closely before giving their approval. In Mr Bumble's case, they could not get close enough to inspect him, so contented themselves with inspecting his house, a large roomy cage in the shape of a caravan hanging on top of an oak stand. Having satisfied themselves that this newcomer to the family proved to be no threat, they showed no further interest in Mr Bumble. He was the most beautiful bird and the biggest of his kind I'd ever seen. We laid him to rest in a shallow grave at the front of the cottage under the lounge window, where the girls never went digging around. The lounge seemed very strange for a while without his cage in the window recess. He was the loveliest shade of green you could imagine and a great loss to us all.

* * * *

I had for some time been concerned about Shelley's weight and finally decided to take her to the vet to try to find out why she was so thin. The question had been bothering me for some time, especially since she was eating well. I was shocked when the vet put her on the scales to discover she weighed only 49 lbs. He could offer no explanation but gave us some tablets and sent us home. We stepped up her food supply and she seemed full of life. So much so that about a week later, she and Samantha, although they had always been close, had their one and only scrap. A marrow bone had been left on the table in the kitchen for later use. I was in the kitchen at the time with the four girls when the two young ones made a grab for the bone at the same time and Samantha turned on Shelley. I managed to separate them but got bitten by mistake for my trouble.

* * * *

The pups were fast approaching their second birthday and, as far as we knew, Shelley had never been in season. We waited and waited but there was no sign and, as Shelley was eating up her food and full of beans, we didn't pursue the reason.

As we lived on top of a park with ponds and lakes, it seemed inevitable that we would have the occasional rat in the garden. One morning, at the end of January, Douglas was on the phone in the reception hall, which looks out on to the front garden, and he shouted to tell me there was a rat - a very large one - running around the garage area.

We phoned the town hall to report the matter and to ask if a vermin control officer would call. When he came he was a nice man who was fond of dogs, yet, to my surprise, he told me that the only sure way of killing rats was with a bait of warfarin mixed with sugar. We were worried about this, as the dogs had the run of the entire garden. I'd been thinking in terms of a cage into which the rat would walk, as with a mouse trap, and which could then be taken away.

A Golden Love Story

After much thought, we agreed to have the mixture put down in an area we thought the dogs never visited - a slipway between the fence which divides us from the park and the garage and car port. A week passed and there was no further sign of the rat.

* * * *

It was the first week in February when we paid our first visit to Crufts Dog Show, arriving home for afternoon tea and the dogs' feeding time. Shelley was off her food and limping, so we took her to the vet that evening. He had no idea what was wrong and a pang of fear went through me as I remembered the warfarin and I mentioned this to him. But the vet thought it was unlikely to be the cause of her troubles, explaining that it would take a massive dose to produce such symptoms. That night, we took her up to our bedroom to sleep on her blanket on the rug by my side of the bed. She didn't stir throughout the night and the next day seemed weaker and much worse. The vet arrived and still couldn't say what the trouble was. He gave her another injection and said he'd be in the next day.

We sat with Shelley until Alfred could take over that evening. Douglas and I had promised to take some friends to a charity dinner in London and we had to keep our word. But, needless to say, it was a dreadful evening for us, and we not only rang home to see how things were but also rang the vet to try to find out the results of a blood test he'd taken.

Shelley was as delighted as ever to see us when we got home, despite her obvious weakness. She couldn't get up but she still managed to wag her tail strongly in greeting. We'd left the dinner early and again carried Shelley up to be with us. By now we were seriously worried and feared the worst.

Shelley was very weak the next morning and a different vet arrived to see her - one in whom I had much more confidence - and he confirmed my worst fear, that it was the rat poison that was making her so ill. There was only one hope. She would have to have a blood transfusion from a near relative, and little sister, Samantha was the obvious choice.

I was desperately anxious, especially since Douglas was very busy and could not help much. He had recently joined the board of directors of the Army & Navy group and had an important meeting he couldn't rearrange. The store had been raided and there were statements to be made to press and police and a post mortem to be held on security arrangements, etc. So who was going to help me with Shelley? We decided that I shouldn't drive the girls in my own car to the vets in Croydon as I was too upset to concentrate properly and wouldn't be able to lift Shelley on to the back seat unaided. We rang around and found a car hire firm willing to take dogs. In due course a kindly driver arrived and lifted Shelley into the car for me. I got in with Samantha, cradling Shelley's head in my lap and did my best to comfort her as we sped towards the surgery. I walked in with Samantha while the vet carried Shelley inside. Then I kissed them both and left, knowing that the transfusion equipment had been loaned to the vet especially for this emergency. I returned home and waited until six, when I phoned the vet to find out how things were going. The transfusion had gone well and Samantha could be collected that evening, I was told. I didn't see Shelley, as she was

Grown-up Goings on

sleeping and would need to be kept in for a day or two. But I was happy to collect Samantha, who seemed none the worse for her experience, and felt hopeful and grateful that at last something was being done for her sister.

I had some dinner and popped across the road to visit friends for an hour. At 10.30 pm the telephone rang and Douglas took the call. It was the vet's mother, who lived at the surgery, to say that Shelley had passed away. She had rallied briefly and appeared to be quite lively, and it was thought that the transfusion had done its job. But then quite suddenly she keeled over and was gone. We just couldn't believe it. We were absolute shattered.

Everyone who has experienced bereavement will know that one of the worst things is waking up the next morning, if sleep has come at all and realising all over again that something terrible has happened. These were my feelings that morning, but one thing was certain in my mind: Shelley must be buried close to us at home.

We were sitting in the breakfast room where she was born when Douglas asked me what we were going to do. He already suspected that I wouldn't want Shelley to be taken from the vets without bringing her home. At last I managed to say:

'I'd really like to bury her at Tamarisk. We've such a large garden there with plenty of room and we could put her in a nice peaceful corner overlooking the sea.'

Douglas agreed. So the next task was to telephone the vet and ask if he'd keep Shelley a little longer, while arrangements were made for a suitable box to be obtained and for someone to collect her. We had to do some fast thinking. Who, of all the people we knew, would make us a box? One name sprang to mind - a kind and helpful man Douglas had known for some time - Ken Banham, who was himself a dog owner. He understood our request and readily agreed to help and, later that day, Shelley came home wrapped in her pale green woollen blanket in the large box which had been quickly made for her.

The next morning I drove down to Kent with Shelley and a handyman named Hugh. He knew the dogs and had come to dig the grave in the spot I'd chosen overlooking the sea and large pond her grandmother had investigated a few years before. It was a lovely mild day for February, with no rain or wind. When our sad task was done, Hugh went off to the pub for a sandwich and a drink but I decided to stay indoors, having no wish for food or company. As I stood in the large bay window in the lounge I suddenly became aware of one of the most beautiful rainbows I'd ever seen arched high over the water. It seemed to last longer than usual and I watched, transfixed, until it disappeared from view. Suddenly, I seemed to feel better. Perhaps everything was well with our darling Shelley after all.

* * * *

Our great friend Gladys had left a red rose to be placed on Shelley's resting place and I did this before driving home. The motorway seemed extra long and a dreadful tiredness swept over me as I drove: the sun was warm on the car for the time of year, which didn't help. Hugh couldn't drive, so I wasn't able to hand the wheel over to him. Instead, I opened all the windows and kept Hugh talking until I'd arrived home without mishap at the end of an emotionally draining day. That evening. I had to chair

A Golden Love Story

the PDSA guild meeting, which was the last thing I wanted to do but, in retrospect, it was a good thing and far better than moping at home.

The country was suffering from power strikes and several events had to be cancelled for the coming week. This was a pity, from my point of view, as it would have helped greatly if I could have been kept fully occupied.

It was strange how big the gap in our lives was with Shelley gone. You might think that, with three dogs still in the house, the absence of a fourth wouldn't seem too great a breach. But every dog is an individual in his or her own right, with a special personality that will be terribly missed. As with any bereavement, there are always those little hurdles and special dates to be approached and overcome.

The first time I returned to the dog training club and watched Shelley's class working I felt very sad, knowing she should still be there and, worst of all, feeling it was my fault that she wasn't. The next hurdle was her second birthday at the end of February and I was relieved that, by the end of the following month, her resting place had been smoothed down ready for planting and a surround of patterned slabs placed around in a neat square patch. Yet, to the day I die, I shall never forgive myself for having warfarin concealed behind the garage and for not spotting the signs that Shelley had ingested it sooner. As I write this page, it happens to be the 14th anniversary of her passing, and as always, I say a special prayer for her.

* * * *

We were staying at Tamarisk for the Easter holiday and, during the afternoon of Good Friday, a couple called on us with some snaps they'd taken of all the girls in the garden the previous summer. They'd enjoyed watching them at play and had asked our permission to take the pictures. We'd completely forgotten about the incident and they insisted we had two copies of the film. Little Shelley was right in the front with Samantha and they'd both come out very well. We were delighted to have them, as they were the last pictures we would ever have of all our little family together.

Chapter 9
SIBLING RIVALRY AND ALL THAT

Sally's daughter, Salvia, came to stay with us for a week while John and Sheila were on holiday. She was a lovely light-coloured golden retriever with a sweet nature. Shandy adored her and loved having her to stay. She went out to greet her when she arrived, escorted her indoors and took her under her wing in general. Sally, however, wasn't quite so thrilled to have one of her children back in the house. She grumbled most of the time, not to Salvia, but to us. She would climb on to our laps crying and softly growling under her breath, with her eyes fixed on Salvia. We found this strange, I must say, as she was quite happy to have an unknown dog or bitch visit us.

Samantha was different again. She was absolutely furious when she saw Salvia in the kitchen for the first time since she'd left home. She chased her sister through the hall and upstairs, where she proceeded to attack her without provocation of any kind. Now it was my turn to be furious and, after rescuing poor Salvia, I grabbed Samantha, shook her hard and bellowed at her, but all to no avail. She was adamant: no dog, not even a sister or brother, who didn't live in the house, would ever cross the threshold during her lifetime. So, for the rest of the week, we had to keep the doors on the ground floor closed, with Shandy and Salvia on one side and Samantha and Sally on the other, not easy when, on some days, there were four people in and out.

The day after Salvia's arrival was a Sunday and we took her with Shandy to Tamarisk, which they loved. They played happily on the beach and in the garden until it was back to Elmwood the following evening.

* * * *

It was obvious that, unless someone was going to be home when I needed to go out, Salvia would have to come with me everywhere. We went to the launderette together as well as to the hairdressers, the chiropodists, to friends for tea and to the dog training club, as well as a music lesson. Salvia behaved beautifully all the while and was a great success wherever she went. She had her last run with her adoring grandmother in Dulwich Park on the Sunday before she went home. John was waiting at his photographic studios in Lewisham to take her home and Salvia went wild with delight to see him. She was greatly missed by us for some time and it made me sad to know that realistically she could never come and stay with us during Samantha's lifetime.

At the end of May, I planted the first lot of flowers on Shelley's resting place, some pink and red begonias which were left over from planting at Elmwood. It's a shady spot and these small bedding plants seem to thrive in such a setting. I hoped very much that they'd do well on her little spot that summer.

* * * *

A Golden Love Story

Samantha was now in the top class at the training club and doing quite well, except that she still refused to retrieve, for some reason. She was still a good ambassador for the club and, one evening, during the dog training session in September, Samantha had her photo taken with brother, Jumbo, and another 'golden' called Barry, to complement an article on the club in a local newspaper.

The following Monday, I too made the local news when I had a car accident on my way to my hairdresser's, David Raven at 'Fredaire' in Lewisham. A wasp flew into the car and, as I put my left hand up to flick it away, the car veered to the left and crashed into a lamp post. Thankfully, it was early afternoon and there was no one on the pavement at the time. I had debated whether to bring Shandy along for the ride, but it was a blessing I decided against it as she would almost certainly have been killed had she been in the passenger seat.

The car was a complete write-off and I was taken to hospital for x-rays and stitches in both knees. The police informed Douglas, who was attending the funeral of one of his directors. His secretary gave him the news and came back to the hospital with him to lend a hand. I was discharged the same evening but could only get in and out of the car with great difficulty. The next problem was keeping the girls off my lap: they simply couldn't understand at first but soon seemed to sense that something was amiss and contented themselves with climbing on other laps or just laying at my feet.

The following day found me with a black eye, a badly bruised arm and ribs. I was in a really bad way. Pulling myself in and out of the armchair was bad enough but getting upstairs to bed was even worse. Thank goodness for a bathroom on the ground floor, that's all I can say! Alfred did all the dog walking for me and Douglas helped with the meals, while Violet was an absolute tower of strength, as were friends who called round to cope with shopping, making tea and snacks as well as my bed, when I was alone in the house. It was over a week before I managed to venture out again.

Douglas drove us all to Epsom Downs the following Sunday and Alfred gave the dogs a good long run while I sat and watched from the car. Two weeks later, I went to stay in Worthing for a week on my own, as Douglas was unable to get away. Hopefully, I would recover from the effects of the accident in such a pleasant place, with friends in and around the area I could visit.

But it was good to be home again after my little break and to take the girls for a run on my own. Mind you, I wasn't ready to try to kneel yet, either indoors or out in the garden. That little trauma was yet to come.

* * * *

Beckenham Dog Training Club's competition came round once more on 18 October and Sally came second in the novice and Test A categories. Both classes were large and I was very pleased with the results. Jumbo, for some reason, did badly, although he was always an excellent dog at work. But this is how it can often go. After all, even top champions have their off days.

We collected Sally's rosettes and diplomas at our annual dance and prize-giving night in December. I called on Bobby and Salvador that morning, as they were moving the next week to Newport in Monmouthshire and I wanted to say good-bye. We all

Sibling Rivalry and All That

kept in touch by letter and Christmas card and a snapshot would arrive from time to time, but it was several years before we actually saw the boys again.

The New Year arrived and proceeded calmly for the first couple of weeks, except that Violet had been off work for that time caring for her aged mother, who lived with her and needed constant attention. Then, one morning, I was very depressed to receive a letter from Violet handing in her notice, and saying that she felt it was unfair of her to stay home any longer, not knowing when she could return to work. Violet had joined us when my mother died and had moved in with us on a three day a week basis. She'd lived in when we were away, welcomed Shandy at eight weeks old, seen two litters of 'goldens' born and raised, grieved with us at Shelley's untimely passing and helped me through my accident. There were the numerous small events that occur in every household that she had shared but, above all, she had loved the dogs and would go to any lengths to help them and Alfred, when I wasn't there. She was in every way a part of the family.

However, the problem had to be faced, and we needed to find someone just as reliable to take Violet's place. A lady I've known for some years, a PDSA supporter whose judgement I trust, introduced me to our present treasure, Mrs Arnold. I interviewed her the following week with some trepidation. After Violet, I couldn't imagine anyone being all the things she'd been to us and the girls. I called on Violet the next day and told her I thought we'd found someone to fit the bill but that the job would still be open should she feel able to return or if Mrs Arnold didn't find us to her liking. She was very pleased with the news and our offer. Happily, Mrs Arnold understood and accepted this tentative arrangement. She joined us at the end of January, and my fears proved completely unfounded, although it took some time to dispel my anxieties. Mrs Arnold had two very young grandsons, Stuart and Philip, and she'd asked permission to bring them to the house during the school holidays. The boys turned out to be two more playmates for our young 'goldens' and they all got on extremely well, the boys joining in our walks and runs in the Crystal Palace grounds in later years.

I had to have a minor operation in March, which meant a four-day stay in hospital and a week's break in Worthing to fully recover. The best part of the week and the best contribution to my recovery, having been under the weather with a throat infection, was the sight of dear Shandy arriving with Douglas to take me home. Once there with all my girls, my return to normal health was very rapid!

* * * *

Summer brought the normal crop of doggy ailments, tummy upsets and skin disorders which had a nasty way of repeating themselves. Sally and Samantha suffered on and off for the rest of the season. Shandy and Sally had a week's holiday with Alfred in Kent and Samantha stayed home with us. Such a highly independent young lady, she never appeared to mind being on her own and simply carried on as if everything were normal. Yet she must have felt the absence of her family inside and, on their return, they all went wild with delight, the welcome going on and on until they all flopped out.

In the autumn, the club obedience competition saw Jumbo winning first prize in the novice class with mum, Sally, getting a second. Samantha lost all her marks by flatly refusing to retrieve the dumb-bell. But, the next week Jumbo won the more difficult Test A and Sally came third. The club moved to a new hall in December and Shandy, now aged nine, and who had long ago given up classes, came along just to watch.

* * * *

Douglas had met a portrait artist named Derek Chittock just before Christmas and they discussed the idea of having my portrait painted with the three dogs. In the event it transpired that such a portrait would take up a large section of the dining room wall and might break the bank into the bargain. So we finally agreed that just Shandy should join me in the picture.

Mr Chittock arrived early in the New Year for the first sitting, but the light in our lounge, with its low beams and leaded lights, didn't lend itself at that time of year to portrait sitting. So we went with Mr Chittock to his home in Chevening, where the conditions in his studio proved ideal. Sittings lasted half an hour and there were six all together. Our major problem was how to keep Shandy still for half an hour or at least interested enough to look up at me in the pose we wanted. My own choice would have been to have her with head resting on my knee and looking out into the room. But Mr Chittock had already used such a pose with his own labrador who had died shortly before and I knew he didn't want to paint another dog so soon in the same pose.

If there was anything to capture and hold our Shandy's attention it was food of some kind. So, at each session, I took along a bag of biscuits which produced the desired effect. But my feet were soaked at the end of each session with her saliva. Some sessions lasted most of the day and we left home in the morning of 5 March for our penultimate session, arriving back at 4.30 pm, feeling as tired as if I'd completed a marathon run. I think Shandy felt the same. But the portrait was coming along nicely. There were quite long breaks between sittings, during which time Mr Chittock worked on my dress, which was black velvet with some detail on the sleeve and the three-string pearl necklace. I was relieved when the sessions were over. Shandy, on the other hand, must have really looked forward to these outings with the bag of biscuits at the end of them.

Chapter 10
SALLY IN THE WARS

It was early in April when I noticed Sally was holding her head to one side and, on inspection, I found a lump in her ear flap, just like one of those weights we used to use to hold down heavy curtains. Off we went to the vet and he diagnosed a haemotoma (a collection of blood in the ear flap). This would have to be drained and an appointment was made for two days later. No food must be given from 4 pm on the day before an anaesthetic and it was no easy thing to feed only two of three dogs, especially ones with appetites and tummy clocks like ours had.

Sally went to the vets early as she needed to stay in for the day. When we collected her late in the afternoon, she was still very dazed, so we kept her quiet on her bed for the rest of the evening. Her ear and head were swathed in bandages. What a sorry sight she was! None the less, we expected her to be full of beans next day after a good night's sleep and not too bothered about those bandages. We kept Sally with us overnight, lying her on the foot of our bed, in case she was distressed and to stop her from knocking her head. She didn't stir all night, or in the morning when we got up and the rest of the family came to life. We were worried enough about her to send for the vet. When the vet arrived he assured us that Sally was simply suffering from an anaesthetic hangover. 'I promise you she's not going to die,' he said. 'By this evening she'll be as lively as a cricket.' And, sure enough, she was.

It was Good Friday next day when we took her bandages off and all appeared to be well. She was also as good as gold throughout. Her stitches had to come out ten days later and she gave us no trouble at all.

The portrait Shandy and I had sat for back in the winter arrived the following week. Douglas was very pleased with it but I found it needed a lot of getting used to. I asked for it to be put on the dining room wall, so that it was behind me when I sat at the table and when I sat in the armchair watching television. I loved gazing at Shandy but I didn't particularly want to see myself quite so frequently.

Tiresomely, Sally's ear blew up again four weeks later so it was back to the vet for draining and more stitching. We again collected her at tea-time and kept her quiet for the rest of the evening. She was off colour the following morning, just as before, but this time we knew what to expect and we let time and nature take its course, without calling in the vet. She was her old self by evening and when we took her back for a check and injection, everything seemed fine.

We left for a holiday in Cornwall the next day, knowing our pets were in good hands at home. Sally had her stitches out when we got back and we just had to keep our fingers crossed for better luck this time.

* * * *

A Golden Love Story

Life continued in much the same way for the girls, with dog training for the younger ones and trips to Tamarisk as often as we could. The one and only time the girls were ever left for a whole day on their own was one Sunday when our friend Gladys came to stay and we all went to see the National Stud at Newmarket. We gave the dogs a good run in the Crystal Palace grounds and left dinners for them, which I suspect were eaten as soon as the front door closed behind us.

Our burglar alarm bell was ringing when we arrived home but there was no sign of anything wrong and the girls were certainly thrilled to see us back. Shortly after this I was honoured during a Beckenham Dog Training Club committee meeting to be invited to become a trainee instructor, and was pleased to accept. The very next night I started my apprenticeship in the first class where it had all started for Shandy and me nine years ago. Little did I dream then that I'd ever be joining the committee or become an instructor myself.

Sally meanwhile was doing well in the top class and took part successfully in one of our demonstrations at a PDSA fête in Sydenham. Unfortunately, she was again in the wars at the end of September. This time a bare patch appeared near the top of her tail which we noticed she couldn't leave alone. The vet discovered that her anal glands were full and this was the reason for the tail chewing and he had to empty them for her. It was a very messy and smelly business: a black foul-smelling liquid was emitted and, what is worse, once these glands start to need emptying they seem always to need the same treatment. Yet our dogs were always given plenty of roughage in the form of cooked green vegetables, raw grated carrot and wholemeal bread, baked in the oven to make rusks. Poor Mandy was afflicted even more seriously and eventually had to have her anal glands removed completely to save her from further distress.

* * * *

Shandy reached double figures on 4 December 1974. She celebrated, as always, with a special treat of a marrow bone, and each member of the family had one to keep her company. Two days before the big event, our portrait went into an exhibition at Farnborough Hospital medical school, at the request of the artist. Three days after Shandy's birthday, Sally, our other 'old timer', was awarded the Veteran Cup at the training club at our annual dance.

When the portrait came back from Farnborough, we had it photographed; I gave a large framed print to Douglas for his birthday which hangs in our lounge on the coast. It seemed a nice idea to have several small ones printed for family and friends as Christmas presents too. Our friends John and Ethel came to dinner early in January, and brought Rusty with them. The girls were delighted to see him, and even his mum, our Sally, seemed pleased. She hadn't been at all happy when her daughter called and looked like staying, as we knew only too well. Or perhaps I'm misjudging her and the difference in the welcome was because Rusty's owners came with him and took him away again at the end of the evening. Sally was back to the vet in March for her other ear to be drained and, although she remained dopey for quite a time after the operation, she wasn't as bad as before. She was back for the first ear to be drained exactly one month later and was a sorry sight that evening, with both ears bandaged. We'd never

Sally in the Wars

before experienced anything like this with our animals. To make things worse, she removed the bandages, and had to be rushed back to the vets yet again. But she did manage to keep them on for a few days and the vet finally removed them to reveal a 'pretty rough' pair of ears. It took plenty of ointment and fresh air for the first ear to begin to improve and, with the healing process now in full swing, we could only hope it would last for good this time.

* * * *

The girls' health was very good on the whole during 1975, apart from the odd skin eruption and minor tummy upsets, so common during the summer months. Shandy showed little sign of arthritis any more, thanks to those magic herbs. But there was a plague of mites during October, and we had to bath all three dogs with a preparation called Alugan, which thankfully did the trick.

Sally was again awarded the Veteran Cup for her work throughout the year, but she'd decided that enough was enough and refused to work to command in my time. When we were standing in class, awaiting our turn to work, she would constantly put her paw up and tap me over and over again on my leg as if to say: 'I'm really fed up with this.' So I retired her from obedience work and she often came to the club as a spectator instead. Shandy, now 11, joined in the doggie party at Beckenham and loved every minute of it.

Shandy had a go in her old top class at the dog club early in 1976 and did very nicely indeed: once trained, dogs never seem to forget the exercises, even if they do have to execute them at a slower pace. I didn't make a habit of taking her at her advanced age, but the odd occasion turned out to be fun for both of us. I was now taking the third class on my own more often, so our Thursday evenings were busy ones.

* * * *

Maybe things were going too smoothly, I don't know. But one Monday morning in May, Mr Madden phoned to ask if we could get Jumbo to the vet as they had no transport. He suspected a blockage, he told us, due to a cooked lamb bone. I was glad to drive them to our vet in Croydon, where Jumbo had some treatment and was booked to return the next day. I wasn't unduly worried at this stage: Jumbo had walked out to the car and jumped in without much effort. But an early phone call next day told us he was much worse. This time we had to carry him out to the car and lay him on the back seat before speeding back to Croydon, where we left him with the vet, going home to await news later in the day. I had an appointment at the hairdressers that afternoon and, although I kept it, Jumbo was uppermost in my mind as I tried to read under the drier. Then Mr Madden rang to say our beloved Jumbo had died during the morning. The Maddens were shattered and so was I. Jumbo had chewed a bone from the Sunday joint, as he had done many times before, but this time a sharp piece had broken off and penetrated the intestine, causing peritonitis. Nothing could have saved him.

I had to go to a dog training committee meeting that evening and everyone there was shocked by the news. They all knew him so well and remembered how good he

A Golden Love Story

was in his various classes. And now, like so many of us, the pictures and trophies standing on the TV were all his owners had to remember him by. Jumbo was six years and three months old.

I kept my engagement at the Luncheon Club in Hayes the following day. I drove from there to pick up the Maddens in Forest Hill and take them to collect Jumbo's body from the vet's. We carried him between us to the car and laid him on the back seat, wrapped in a blanket, covered by a black plastic bag. The journey was a distressing one and sad in the extreme. We carried Jumbo into the Maddens' sitting room and laid him out on a blanket in front of the fireplace. He looked as if he was sleeping so peacefully that I could hardly believe he'd gone. The Maddens planned to bury him in their own garden.

I called at the house two days later to see them, as well as the spot where Jumbo had been laid to rest. Mrs Madden said the body had been quite wet from her husband's tears as he knelt over him, covering him snugly with his big blanket for all time. That dog had been their entire world, as Mrs Madden said: 'We've loved all our dogs and were heartbroken when each one had to go. But our Jummy was something very special.' They have never really recovered from his untimely and unnecessary death to this day. Going to the fish shop for the next few months was not a trip we relished: it seemed so strange not to see Jumbo sitting there watching us and the world in general with great interest. George invariably had a tale to tell about Jumbo's exploits and he was now so quiet and thoughtful. It was still early days and time would eventually heal the wound, although the scar would always remain.

I related the whole story to our friend, Holland, and he listened nodding wisely several times. He said: 'Phyllis, over the past thirty years I have had not dozens, but hundreds of dogs brought to me with the same problem and almost every case has proved to be fatal. Nothing can be done once peritonitis has set in.'

Buster Lloyd-Jones also touched on the same subject in one of his books on herbal medicine, implying that some dogs can chew bones successfully while others can't. So, from then on, we decided we would never try to find out with pets of our own.

* * * *

The summer of '76 was hot and, for my part, I hate the heat and can't bear to think of dogs becoming miserable in their heavy coats. One day in June it went over 95 degrees and even when the girls went out at 8 pm it was still very sticky. We tried to overcome this when they were in the house by leaving the doors to the garden and all internal doors open in line with the double front doors, so that a current of air blew through the house. We kept the front doors closed and also kept a sharp eye on the front in case anyone called and left the gates open.

It was so hot that I took the girls down to the beach with some friends when we visited Tamarisk at the end of the month. Our friends had a swim while I had a paddle and, to our great surprise, Shandy ran in the water with her daughter and granddaughter. What absolute bliss!

Chapter 11
CANADIAN CAPER

The summer continued to be very hot and, in July, we went to Canada for three weeks, mainly for the Olympic Games and also to visit an elderly doctor and his daughter, whom we'd befriended during our trip to South Africa.

Animals were all around us in the lovely resort of Moraine Lake and we met a splendid St Bernard, aged six and a half, called Sandy as well as hundreds of chipmunks who swarmed around for nuts and other goodies. The girls went mad on our return and were so happy that things were back to normal. Samantha was promoted to the top class at the club, despite the fact that she would do only half a retrieve. The powers that be, seemed to think that if they put her up regardless she would change her mind. But she never did, bless her heart.

Soon after our return from Canada, we were staying at Tamarisk with Shandy and a friend named Pat, who had two white toy poodles. We'd taken the dogs for a run on the prom before dinner and, during the meal, noticed that Shandy was not lying near the dining table, as she always did, and indeed was nowhere to be found in the bungalow.

It was so hot, I guessed she was somewhere in the garden in a cool spot, but I was wrong. Panic now started to take over, I must confess. One of the gates was ajar and she must have slipped out. It was getting dark as I ran along the front calling her name for quite a way. But no sign of her. I was very worried indeed by now and went to the local police to report her missing. It was two hours since we'd noticed she was missing and it was so unlike her. Meanwhile Douglas and Pat went along the front in the car, one driving and looking out one side and the other scanning the beach on the other side. It was almost dark and I decided to wait at home and watch from the window. I didn't have long to wait. As the car drew level with the fence I could just make out Shandy's silhouette, sitting in her seat at the front. The feeling of relief was overwhelming. Douglas had found her at a place called Swalecliffe, about two miles away, just sitting there, minding her own business.

She had a good drink once indoors, and then settled down to sleep. Like us, she was worn out but the question remains, why? Why would a dog who hated us to disappear out of her sight stay away from home for over two hours? Was it because she resented the presence of Little Lady and Madam, the two poodles occupying her house and who didn't look like leaving? Or did she simply fancy a walk without them? Perhaps she was even looking for Sally and Samantha, who didn't come away with us on this rare occasion. We'll never know. But at least now she was home safe and sound and that was all that mattered.

* * * *

A Golden Love Story

The weather was cooler now, which was good, as the following Saturday Samantha took part in another obedience demonstration with Beckenham Dog Training Club at a garden fête which I'd organised for the PDSA in Sydenham. We were grateful for the cooler weather, but could have done without the downpour just after our display which closed the show by 4 pm, but not before we had raised £177 for the society.

In September, all the girls did a sponsored walk around the lake area of Crystal Palace and Rusty came too. It was beautiful weather and about fifty people walked with their dogs in aid of the PDSA.

I decided not to work Sally in the club's yearly competitions and returned the Bromley Trophy she'd been holding as best veteran. Shandy had her twelfth birthday on 4 December and, by way of celebration, the girls had a marrow bone each, which they gnawed on for some hours. Little did we know the devastating results that Samantha would undergo as a result.

Since the tragic loss of Jumbo, cooked bones were out, but we continued to give the dogs a marrow bone on special occasions. We'd been to Sussex for lunch with friends and, on our return, it was clear that Samantha was in some distress and was unable to pass anything. We wasted no time in taking her to the vet and he said she had constipation due to crunching the bones. The next morning, Douglas took her for x-rays and an enema and she stayed with the vet for a while before he brought her home. But she was still in distress, to our great dismay, when the 'dope' had worn off. We took her back to the vet twice the next day, and another x-ray showed that her bowels were still not clear. She seemed brighter and more relaxed that evening, but I still felt anxious so it was back to the vet again next morning. There was still no joy and I was feeling quite worried now. I went to town with a friend to see a new show called *A Chorus Line,* but at the back of my mind was Samantha and I wondered constantly what the news would be on my arrival home.

It was not good, I'm afraid, and it was back to the vet's the following morning, where I told him in no uncertain manner that this time something positive must be done, no matter what the cost. He kept her in for yet another enema and when I called to pick her up, she came out on to a bit of rough ground and at long last passed everything that had refused to budge for a week. What a great relief all round! She had a check-up the following day and an injection against any infection that could have been lurking and our Samantha was out of the woods, at last. Who'd have guessed that the celebration munching of those marrow bones would have turned out so badly? True, we'd had the sad episode of Jumbo with the cooked lamb bone, but we'd never given these to the girls and, from now on, would never give another marrow bone, uncooked or not. Some dogs seem to manage any bone but we were taking no more chances. Samantha was put on cooked fish and chicken for a day or two and then back to a normal diet until she was her old self again.

She went to the vet for a final check-up and all was well. The bill came to £32.50, but we made no bones about it. It was worth every penny to undo the damage done to Samantha. She came with us to the coast that day and flew along the promenade without a care in the world.

* * * *

Douglas, chairman at the PDSA Exemption Show, Crystal Palace Grounds, now known as the Douglas Seymour Memorial Dog Show. (Photo: John Brown)

The 'Sunspray' Gang, aged 4 weeks. (Photo: John Brown)

The 'Sunspray' Gang, aged 4 weeks. (Photo: John Brown)

Shandy, aged about 4 years.

Shandy in later life.

The whole family in the garden at Sydenham. Puppies aged 12 weeks. (Photo: John Brown)

Shandy and Sally around the Christmas tree, 1967. (Photo: John Brown)

Rusty at his home in John and Ethel Gregory's garden. (Photo: John & Ethel Gregory)

Alfred with Shandy, Sally and Samantha, Simon Groom and Goldie from Blue Peter.

PS with Douglas, at the first PDSA Exemption Show in Crystal Palace Grounds. (Photo: John Brown)

The portrait with Shandy by Derek Chittock. (Photo: John Brown)

The miniature of Shandy, Sally and Samantha painted by Elizabeth Wood. (Photo: John Brown)

The very last picture taken of all the girls together.
(Photo: An unknown couple of well-wishers from London)

*PS with Scott and Selina (present dogs) in the garden at home.
(Photo: Peter Jempson)*

The Sunspray Cup, presented to the Southern Golden Retriever Society in memory of Shandy. (Photo: Southern Golden Retriever Society)

Burglars came again in April, despite the fact that the girls were indoors and an alarm now in operation, and we were only away on a trip to Brighton for a few hours. We were thankful that the dogs were unharmed and no damage done but the experience was no less upsetting than before.

Jumbo had been gone for a year on 18 May and I called on the Maddens to see Fred, their new 'golden' puppy, who was adorable, and to see Jumbo's resting place in a quiet corner of the garden. It was a mass of bloom, with not a weed in sight. The Maddens were mourning still, but Fred was a great help to them and I think they were very wise to have another puppy at that stage.

* * * *

In October, I went alone to Australia, stopping over in Hong Kong on the way out for a few days. Douglas drove me to London Airport and Shandy came to see me off. It was a long three weeks on the other side of the world and, despite a happy stay with an old school friend I hadn't seen for thirty years, I missed home and the girls very much. I was due to go to Queensland to see the Great Barrier Reef and then on to Thailand, but decided to cut my trip short and return home.

Shandy and Sally came to the airport with Douglas to meet me but the plane was late and we were so long disembarking that they were all well and truly fed up when we got back to the car. Had it been summer it would have been even worse as Douglas would have had to take the dogs home and come back for me. Samantha, who'd stayed at home with Alfred, gave me a great welcome and it felt so good to be home with everyone again. The flight from Sydney seems never-ending and one's body clock goes haywire for a few hours. So I retired to bed at 9.30, with the girls as happy as I was that 'mum' was back.

The following day was Douglas's fifty-ninth birthday and we took the girls to the park before going for a celebration meal in Westerham. Samantha was now also on the herbs we gave Shandy, who was now 13 and leading a reasonable life, as far as we could tell. Certainly, she still enjoyed her visit to the annual doggie party that December.

Dear Shandy won the veteran class at the PDSA Exemption Show in March; Alfred took her in for us and proudly came out with the red rosette and cup. I was elected chairman of the Beckenham Dog Training Club, an office I was to hold for nine years until I had the honour of becoming president in 1987.

The summer of '78 was a cold and cheerless one, so it was rather strange that the girls should have an outbreak of skin trouble normally associated with the heat. Injections soon did the trick and, in September, Sally took fourth prize in the veteran class and was looking in pretty good condition, as befits this particular class.

* * * *

Our elderly ladies had long since passed the days of taking part in obedience demonstrations, but I was still doing my stint commentating and our club was invited to give a demo at the Guildhall as part of a programme of events organised by London taxi drivers in aid of under-privileged children.

A Golden Love Story

I was pleased to be able to assist but wistfully thought, as I watched the dogs going through their paces, of our old 'uns at home and how eagerly they would have taken part in these events in earlier years. The following week we'd tried to book a holiday in Suffolk without success, so we settled for a break at Tamarisk and this meant that the girls could come with us.

Once the family had been out for a good run they would settle down for a few hours indoors until we returned from a trip out to give them an evening meal. After a good roam around the garden and sometimes a second run along the prom, they were ready to relax in readiness for the evening, leaving us free to go out for a meal or to enjoy an evening of television or reading.

For me, it was often a case of cutting tags for Christmas gifts from old cards to sell in aid of the PDSA and I still do this today. There is always a big demand for them and it's very gratifying to see a finished product from something that would have been thrown away and which regularly brings in £75-£100 for the charity.

A Croydon newspaper phoned on our return to ask if they could do a write-up on our forthcoming dog show and sponsored walk at Crystal Palace. They needed a picture of my retrievers and me on a walk! This was taken on the forecourt of the PDSA and it made the front page the following week. How about that for a bit of instant fame and glory? Our annual dog show and sponsored dog walk came round in October. Our three old ladies once again took part, albeit a small one, and enjoyed lying on the ground watching the younger, more active dogs walking and working. This year we were lucky enough to have a real canine celebrity in the shape of Goldie from *Blue Peter,* who came with presenter Simon Groom. She was a most appealing retriever and walked a lap for the PDSA as well as taking part in the obedience display. It was a warm sunny day, ideal for taking photographs and Goldie happily posed with our three for some shots.

* * * *

I arrived home from the hairdresser's two days later to find Shandy clearly unwell and distressed. She was roaming around the house panting and drinking copiously. On closer inspection, I found a discharge from the rear which was coming from the womb. We rushed her to the vet, who confirmed she was very ill indeed. He explained she would need an operation to remove the womb, which had suddenly swelled up - a condition know as pyometra. The vet warned me this was a serious condition and the risk of an operation for a dog of Shandy's age was obviously greater than for a young dog. But without it she would surely die, so the risk had to be taken and Shandy was sent home overnight after an injection. I stayed downstairs to look after her during the night. But we had little sleep. Morning seemed a lifetime coming round but, when it did, we took Shandy in for her operation.

It was a warm, sunny day, almost like summer, and I kept as busy as possible. We rang the surgery at 4 pm to be told that all was well so far and we could pick the patient up quite soon. Alfred came with me to help carry Shandy, wrapped in a blanket. She was very sleepy and needed to be kept nice and quiet and we all sat down to a good dinner with her sleeping peacefully nearby. Alfred stayed downstairs with her that night and I took over next morning. She had a meal and went into the garden for a wander round. I stayed home all day to keep an eye on her and stop the others from worrying her. Douglas was away in Burgess Hill, so Alfred came with me when I took Shandy for a check-up. 'So far, so good,' said the vet, but it was still early days following such major surgery on such an old lady. Still, she was winning. Our friend, Holland, paid a social call the following day and had a look at Shandy. We were delighted when he told us she was doing splendidly. I stayed downstairs with her for another night and next day she had an injection of vitamin B12. Our vet was very pleased with her progress and she could now come upstairs to sleep where we could keep an eye and ear on her until fully recovered. Shandy was out of the woods and we were able to leave her with her family and Alfred while we visited friends in Buckinghamshire the following Sunday. They had just got a new 'golden' puppy, called Cash, and she was pale golden and very sweet. Their daughter Jackie worked in a bank so the name seemed appropriate.

Shandy had her stitches out and was so good about this, enjoying the fuss when it was all over. Sally went along too, to have those anal glands cleaned once again. The bill for all this in 1978, including Shandy's operation, injections, visits for check-ups and Sally's treatment, came to £34.78. We thought this was very reasonable and worth every penny. In fact, we would have paid treble or more to get our Shandy well. She was certainly fine now and we prayed she would go from strength to strength. We took her and Sally to the coast for the weekend. The following day saw fine October weather and Douglas worked in the garden and cleaned the windows. We had a quiet afternoon and evening, watching TV and Douglas roasted some chestnuts from our tree at Elmwood. We had a nice light supper of tomatoes on toast, after which Douglas complained of pains in the chest which he thought must be indigestion. After a sleepless night, Douglas was still in pain, so I told him I'd drive him to the local doctor, as I suspected this was more serious than indigestion. I was right. The doctor sent him home to await a nurse who came to give him a cardiograph which showed he'd suffered from a small heart attack.

After a better night's sleep, Douglas suggested I take Sally back to Elmwood. We went by train to Bromley where we were met by Louise, who had acted as a chauffeur for us many times. Sally and Samantha were to be looked after by Alfred and Mrs Arnold whilst I took enough clothes back to Tamarisk for two weeks. I made a lot of phone calls for Douglas and planted our bulbs in the front garden. I was helped by Mr Hobden who was working for us and who would follow me around with chicken wire to keep the squirrels from eating all the bulbs. Then after a hectic dash to the bank and the library, I kissed the girls farewell and was seen off by Mrs Arnold and Mr Hobden at the gate as Louise prepared to speed me off to Bromley for the train back to Kent.

Dr Horan called in at tea-time and was pleased with Douglas's progress. The blood tests were good and we simply had to jog along quietly for the next two weeks. Douglas and Shandy had both given us a nasty fright but they were providing ideal company for each other. I would take the old lady for her run on the prom in the morning and again at tea-time and, the rest of the time, she would sit or lie by Douglas, being fussed or just dozing.

A Golden Love Story

I made the trip back to Elmwood after another week to pick up post, collect clothing and to see to things in general. It was good to see Sally and Samantha and note how well they were being cared for while I was away. The weather was surprisingly good for the time of year, which was a great help. Shandy's feet were clean when she came in from her runs, as it was dry outside, and also her undercarriage, which was now lower than in her early years, still had a good deal of feathering which would have picked up the mud in seconds.

* * * *

Douglas maintained his steady progress and the week of 5 November turned out to be warm and sunny. He came round the shops for the first time since being taken ill and the next morning took Shandy along the front for a short walk in the sun before leaving for Elmwood. Louise drove us back, with Douglas and Shandy both sleeping in the back seat. Shandy had been my constant companion during those two weeks, walking, going by car to the shops and lying at my feet while I painted three gates for Douglas around the garden.

When we arrived home, the girls went wild with pleasure at the sight of each other and us. At the dog club the next morning everyone was very kind. I received a great welcome back and many good wishes to Douglas for a full recovery. Our own doctor called in to see him and was pleased with his general condition. Life returned to near normal, except that Sally and Samantha had to go to the vet. Samantha wasn't eating, so she had an injection and was found to need her teeth scaling.

The following Saturday was 18 November and Douglas's sixtieth birthday. We celebrated by going to a favourite restaurant in Godstone for lunch. The doctor called in the afternoon and pronounced all was well with Douglas and, that night, I attended the Beckenham Dog Training Club dance, presenting the annual awards and other prizes for the first time as chairman. We began to get out and about again, visiting friends and making shopping trips. Douglas wasn't driving yet and I did the honours for a while, much against his wishes, as he liked to drive himself by daylight, though he wasn't so keen in the dark. Samantha had her teeth scaled the following week and I collected her at 5 pm. The next day she came with us to the park and ran around, full of beans. On Shandy's fourteenth birthday, we took her for a ride to Gillingham, where we had an appointment with some builders who made a great fuss of her. She was in really fine fettle for a dog of her advanced years. This year, Sally represented the family at the doggie Christmas party, one of the few veterans to attend. We were fast approaching the end of a year of mixed fortunes, but it had at least been a fairly good one for the girls. At their age you learn to be thankful for every blessing and bonus.

Chapter 12
MEMORIES OF SAM

As honorary secretary to Bromley Women's Luncheon Club, I had been asked to book the world famous medium, Mrs Doris Fisher Stokes, to speak to us in October. The booking was done on the telephone and we'd never met before. When I asked if she'd give me a sitting out of sheer curiosity, on the spur of the moment, she said yes and we fixed a date. So, on 23 February 1978, I went to see her. Mrs Stokes didn't know my address or anyone who knew me. She simply knew my surname and that was all.

The day of the sitting arrived and I set off to Fulham where she lived, very sceptical but nevertheless keeping an open mind. I was greeted by Mrs Stokes, who was a plump, grey-haired, motherly woman, very bright and cheerful, and she took me into a sunny little room with pictures, flowers and bric a brac everywhere. After a little chat about the weather and the fact that she was just recovering from flu, the sitting began. I hadn't given her a scrap of information about myself and there was nothing about me to suggest I owned dogs.

With the hour-long sitting about half way through, I was reflecting on what a truly amazing experience it had been so far when Mrs Stokes suddenly said: 'You have a dog in the spirit world, don't you, lovey?'

'Yes,' I replied.

'Can you give me the name?' Mrs Stokes said: 'They say they have your dog with them and they are looking after it for you. The name begins with S and ends in the sound "ee". But I cannot get the full name clearly.' I didn't tell her the name was Shelley and the sitting progressed on other lines.

Then she suddenly said: 'Who is Sam? Sam belongs... Sam lives in your house.' I nodded, saying nothing. Mrs Stokes began to laugh: 'They say she is a real star turn. She is a star turn with the name Sam? I think there is some mistake. No, wait a minute. Sam is your dog and that is a short version of her name. She is a great character, can be very naughty and leads you quite a dance in the park.' All of which was perfectly true of Samantha. But there was no mention of Shandy or Sally. Mrs Stokes said 'they' (members of my family who'd already passed on) often came to the park with us on our walks and had many a laugh at Sam and her antics.

The sitting came to a close after some further startling revelations and, for a long time afterwards I found it difficult to take it all in. But, on our walks in the park even now, I sometimes smile to myself and wonder if our unseen companions are enjoying the antics in which most young dogs indulge.

Why is it that dogs, like children, will go along with nothing wrong with them until holiday time comes around? This seemed to be often the case with us and our trip to the Holy Land in 1979 was no exception. Just before we were due to leave, both

A Golden Love Story

Shandy and Sally appeared to be unwell and had to visit the vet. Sally went twice in one day due to sickness but, after a few days following an injection, tablets and a final check-up, all seemed to be well and we set off for two weeks, with fingers crossed that it would remain so while we were away. It did, which was especially fortunate bearing in mind their advancing years. We'd already had a few trips to the vets with Shandy early in the year for ear trouble and a bad foot, which we think she may have injured in the thick snow, the winter having been a hard and snowy one.

* * * *

One hot July weekend in '79, we went down on a Friday afternoon to Tamarisk with Shandy and Sally, leaving Samantha with Alfred for the weekend. I phoned the house to check all was well on Saturday but this time it seemed it was not. Samantha wasn't eating and, feeling disturbed by this news, Douglas insisted on driving home the following morning to bring her back to Kent. That way we could see what was happening. He set off first thing and was back with Samantha by 1.30. She seemed lively enough and bounced out of the car to greet her mother, grandmother and me. I took her to the local vet the following morning, where she had a blood test and a general check-up. She was, by now, eating again.

We had to call at the vets the next morning for the test result. Douglas went round for me and returned with bad news. The blood test showed that Samantha was very ill. The vet had a sophisticated machine which showed the blood count to be so low that it didn't register. It was quite a shock for Douglas.

'Am I to go back to my wife and tell her this dog is dying?' he asked.

'I'm afraid it looks very much like it from the reading,' came the reply.

Douglas seemed a long time getting home from the vets, yet he was only round the corner. When he did arrive, I could barely take in the news. I took Samantha and Sally along the sea front and Sam was as lively as a cricket again and went with Sally into the sea for a swim. I can clearly remember walking along the front that lovely sunny day, 17 July, which would also have been my mother's birthday, and seeing the familiar scene through tearful eyes. All the time I was praying: 'Please God, let her be an old lady like Shandy.'

We returned to Elmwood and I took Samantha to our own vet with the result of the test on a large sheet of paper resembling ticker tape. A baffled vet announced: 'It looks like something out of the space age.'

Then, after carefully examining it, he explained they suspected a growth on the liver. Again, I prayed they were wrong. Samantha was so lively and was eating well again. It all seemed like a bad dream. We agreed with the vet that it would be best to leave well alone and just keep a watch for the time being. I took Samantha back for a check a week or so later and all appeared to be satisfactory. The next few weeks were hot and stormy and brought the normal skin and ear troubles to both the dogs, but they were quickly treated and signed off.

We had a week away in Suffolk, leaving the girls with Mrs Arnold and Alfred. Everything was fine when we got back except for Samantha having a bad foot, which soon sorted itself out. The year continued much as any other, except that we were ever watchful of Samantha, and Shandy no longer felt like coming on long walks with us along the sea front or through the Crystal Palace grounds. She still came for short walks but we realised it was becoming increasingly difficult for her to keep up with us and her family. She was fifteen years old, a wonderful age for a dog by any standards, especially one who'd suffered from arthritis since she was six. By now, as might well be expected, she also suffered with a 'dicky' heart. But she still seemed to enjoy life, within limitations, especially food, trips out in the car, visits to her seaside home and, above all, us. She still gave us that wonderful greeting on our return from anywhere but we didn't go far afield these days, having tired of long distance travel and not wanting to be too far away from the dogs now that they were growing older and might need us in a hurry.

* * * *

The year 1980 began depressingly when I was called to the aid of an old couple who lived down the hill in Penge and, being ill themselves, wanted to have their dear old dog, Bingo, put to sleep. He was nineteen and a half years old and his owners only had a pension and very little else to live on, so I had to ask the PDSA to perform this sad task for them. I took Bingo to them myself, signing the necessary papers on their behalf. It was a disturbing mission but absolutely right for all concerned, especially Bingo himself. He was wonderful for his age but had really had enough of life. Douglas drove us to Croydon so that I could look after and comfort Bingo in the back of the car on his last journey. I was dreading the moment we pulled into the car park and had to take the old dog into reception. I signed the papers, hugged him close and handed him over to the vet. Then I left with a great sense of relief, confident in the knowledge that it would only be a matter of time before his doting master joined him.

The Beckenham Dog Training Club had hired a coach to go to Crufts in February and I went with them. It is always a tiring day but, in the evening, we took Sally to the vet. It was a visit I could have done without, after a long day at Earls Court but, as things turned out, it was overdue. Sally began to have the same symptoms of womb trouble as Shandy, to a lesser extent, but they started to give concern and she seemed distressed. The vet booked her in for the following Monday to have a hysterectomy and gave her an injection to tide her over the weekend. We left her quietly at home with Samantha the following day, and took Shandy out to our friends in Buckinghamshire and then on to Hampshire to see a race horse in which for the first time we had bought a share. We met the horse's trainer, John Holt. The animal's name was Eagle Quest and we were to have quite a bit of fun following her fortunes, or the lack of them, over the next year or two. But my mind was still on the dogs at home, wondering if one or other of them was unwell and I was glad to get back that evening to find Sally quiet and fairly comfortable, as far as we could tell.

We took her in for her operation first thing the next morning and prayed all would be well, particularly in view of her age. I kept as busy as possible for the next few hours, did some shopping on the way home, did some washing-up: anything to help pass the time. We phoned the vet at four to be told everything was fine and Sally could leave for home. Alfred came with us to collect her, as she was going to need lifting.

A Golden Love Story

The vet told us that the operation itself had gone to plan but he had also found and removed two tumours, which appeared at the time to be a blessing. We couldn't have guessed at the outcome months ahead but, for now, she was home and resting quietly. Alfred stayed downstairs with her for the night and, next morning, she seemed to be fine and had a light meal of fish. We took her back for a check-up and the vet was very pleased with her progress. Sally stayed on her rice and fish diet for several days and recovered quickly. We took turns in staying down with her at night until we were satisfied she was out of trouble. During the day she was great but at night she was very restless. But this seemed to pass and she appeared to have recovered completely. Her stitches were removed and she was back to her normal self.

* * * *

Samantha went into double figures that February - in human terms she'd reached the ripe old age of three score years and ten - a good age for a dog. Like us, much can be done to help them attain this age and beyond. If they are fit and happy, that's wonderful. But to prolong a dog's life whatever the cost, to drag out a life in poor health is cruel, to say the least. It can only satisfy the selfishness of the owner to do such a thing. Samantha came third in a large veteran class at a show a few weeks after her birthday. May saw Sally's twelfth birthday, which fell on Bank Holiday Monday, just like the day she was born, and we celebrated together at Tamarisk.

Chapter 13
RIGHT TO THE HEART OF IT

The summer of 1980 was just around the corner when we returned to Elmwood for a week before taking another two-week break at Tamarisk, when the weather turned better. It was on this visit, while it was sunny and hot, that Shandy had her first heart attack. We'd been out early along the sea front for a short walk. All Shandy's walks were by now short, as she panted a great deal during any kind of exertion. We'd just got back to the bit of green in front of the house when she suddenly collapsed. I dashed into the garden where Douglas was working, calling: 'Come quickly, I think Shandy's had a heart attack.'

Douglas dropped everything and rushed back with me to find her still on the ground and very distressed. I jumped into the car and raced round to get the vet. When I got back I found she'd recovered sufficiently to walk slowly with Douglas into the kitchen to lie on the floor. Two friends from the dog club, who now lived in the area, dropped in unexpectedly to see us and were able to sit with me to wait for the vet. He confirmed that Shandy had indeed had a heart attack and took her back in the ambulance for an x-ray and other tests. We picked her up later that evening and she was much more relaxed and contented, though she had to be kept very quiet.

The vet found her heart was enlarged, hence all the panting, but the full x-ray results weren't ready and we had to report back to him on the Monday. The next day, Shandy was much better and more like her old self. But we took no chances and kept her as quiet as possible. Our appointment was for the afternoon and Shandy seemed fine by then. But the x-rays were really bad and her heart was so enlarged, the vet shuddered when he looked at them. We came home armed with a big supply of pink and blue tablets and a letter for the vet back home. I knew Shandy's life had to be kept as quiet and uneventful as possible, with no undue excitement at all. And, deep down, I also knew that time was getting short. The quality of her life was the only thing that really mattered now.

* * * *

We continued much as normal, with only short and gentle exercise for our old lady. But she came on a drive with us to visit friends in Hextable. She'd always loved the car and we hoped the effort of jumping into it and the general excitement wouldn't be too much for her. She coped very well and seemed to enjoy her outing as much as ever. We took her to the vet in Croydon the following week to put him in the picture and also took Sally, who needed an injection for a lump she'd developed. The vet treated this as rheumatism, mistakenly, as it sadly turned out.

The next day was my luncheon club meeting and our speaker was the artist Elizabeth Wood. She gave a talk on her work as a miniaturist and brought a display

A Golden Love Story

along for us to see. It was truly remarkable stuff and I took one of her cards, thinking how nice it would be to have the girls done in oils on ivory.

Sally began to limp very badly two days later, and it was clear that evening that she wasn't at all well. We were very concerned about her and called the vet in next morning. This time he knew it wasn't rheumatism. He gave her an injection and asked if we could get her to the surgery later in the day. The diagnosis was heart trouble. She had three injections and we all carried her in and out of the surgery. She hadn't eaten for two days and we left her to lie quietly in the cool of the hall, with Douglas keeping an eye on her while I went to a fête in Sydenham where we had a PDSA stall. I did brisk business on the Pomagne stall but during the whole afternoon could only think of Sally and wonder how I'd find her when I got home.

She hadn't moved and, although this was hardly surprising as the injections were meant to keep her calm, she seemed much too calm for my liking! The following morning, however, she appeared greatly improved and pottered around the garden. I drove Alfred to Brighton for a holiday and on my return was delighted to learn that Sally had eaten a meal and was almost her old self. We took her for a check-up the following morning. She was indeed much improved but not yet right, the vet confirmed. Next, we took her back for a blood test and she came with me to the park for a run for the first time in days. A few days later, we took her to a vet in West Wickham, where an ECG had been arranged and the results, as well as those from the blood test, were good, apart from a slow heart beat. So it looked as if there might be no heart trouble after all, for which we were truly grateful.

* * * *

After days of rain and cloud the weather brightened up on 8 July, which was the twentieth anniversary of my mother's death. Every year I visit the Streatham Crematorium to see the Book of Remembrance and the inscription I'd inserted in it back in 1960. I sit for a while in the lovely garden where her ashes were scattered and think back over the years. There were so many events to remember in the intervening two decades. How interested she would have been in our 'golden ladies' and how even more spoiled they'd have been if my father, whom I always called Pupsi, and my grandfather had been there. They had both adored animals, with a special feeling for dogs and horses. My mother was fond of animals but, quite unlike us, it wasn't in her nature to get down on the floor and play with them, or have them on her lap to kiss and stroke. I, on the other hand, had always hoped that one day I'd have a dog to treat in that affectionate kind of way.

That day Douglas decided to take our two old ladies to the Crystal Palace grounds for a short run, while I stayed behind to do a job of some sort. I was just preparing to leave for Streatham when there was a knock at the door and I opened it to find a nice young woman I came to know as Tilley standing there. She announced that my husband had asked her to let me know that Shandy had suffered another heart attack, albeit a mild one, and could I please go to the park as soon as possible.

I thanked her and ran to the grounds where I saw the trio; Sally looking bewildered, Shandy lying on the ground and Douglas in conversation with one of the park gardeners who had rushed over to see if he could help. Shandy seemed much brighter when I reached the party, but the burning question was how to get her home. It was a short walk but obviously too far for her in that condition. The man chatting to Douglas had an idea: he would line his wheelbarrow with newspaper and a bit of clean sacking and we'd lift Shandy into it and wheel her home. We must have made a strange sight, pushing a dog in a wheelbarrow, who was now sitting up and wondering what on earth was going on, with another one trotting to heel and eyeing this odd contraption in which her mother was now sitting in state.

We were tremendously grateful to the gardener but the only thank-you he would accept was a cup of tea and he was as delighted at Shandy's quick recovery as we were. I was able to carry on to Streatham for my pilgrimage before the book closed that afternoon. On my return we took both dogs to the vet for a check-up - so far, so good.

When you have dogs and are involved in an organisation such as the PDSA, you are contacted on almost anything to do with animals, from seeking advice to pleas for help. One such plea came on the telephone one Sunday lunchtime just as we were sitting down to eat from one of our very strong supporters, an old lady called Miss French who, for many years, had done much for the welfare of animals in need. This time it concerned an old cross-collie type bitch living in Downham on an estate with an Irish family which, according to Miss French, was in a very sorry state indeed and should have been put to sleep a long time ago. She pleaded with me to go as soon as possible and I promised I'd go as soon as we'd finished eating.

I gathered that the father of the family was unemployed and money was in short supply.

'Does he work at all?' I enquired.

'Oh yes,' said Miss French bitterly, 'with his right arm; he's very good with pints of beer on the end of it.'

She gave me no idea how big the dog was, and a cross can mean anything. Douglas suggested that I take a 'doggie' friend along but I refused his help, knowing he couldn't lift anything heavy and phoned the breeder of one of our present 'goldens', Peggy Stevens. She had many years' experience of 'dogdom' behind her and would know what to do. Peggy agreed to come, and I alerted a vet who said he'd be waiting for us, if necessary. Then I jumped in the car to collect Peggy and sped to the house in Downham.

We found a very sorry sight. The collie was a 12 year-old lady and her stomach was covered in tumours, which were bleeding. Her name was Lassie and she'd been much loved and cared for in all other respects. It wouldn't do to try and guess how long this situation had been going on, but it had clearly been far too long, and we managed to persuade her owners that she must be put to rest.

The husband asked about the fee to put her down and I asked if their two grown-up, working children would like to foot the bill. I was told they would but couldn't manage it that day, and that they'd need a bit of time. The main thing was to get Lassie to the vet as soon as possible and put her out of her misery, so I agreed to pay, and the family assured me that I'd get the money back. I was doubtful about that but at least we'd got permission to take Lassie away and we were thankful for that. The father carried her to the car, as she wouldn't let a stranger touch her, and laid her on the back seat. Peggy sat with her as she rested on a blanket and I drove to the vet in Beckenham. Peggy waited with Lassie while I went in to announce her arrival and tell the vet he'd need to come to the car with a muzzle if he was to carry her out. This he did and we waited in the reception room till poor old Lassie had been put to sleep and then settled the account. It was to Michael Hawken and Clinton Jefferies at this practice in Foxgrove Road that we were soon to turn for help. We got a receipt to send to the owners to prove the job had been done and, I must confess, I didn't expect to hear from them again. But, within a month, they had paid the cost in full, just as the husband had promised.

* * * *

Thursday 31 July was a warm day and we decided to take Shandy to the dog training club that evening, just to watch. She was always interested in our Thursday evening excursions and was extra keen to come out with us. I hadn't taken her to the park during the day, due to the heat, and in the cool of the evening, with only a short car ride ahead, it seemed a good idea to give her a break. Needless to say, she loved every moment; watching all the dogs and the people, as well as the great fuss that was made of her. It was to be her last visit. The following day was again very warm and she seemed off colour but we left her lying on the rug in the cool of the reception hall and went shopping to Lewisham.

We arrived home at lunch-time as John arrived to take photographs of all three girls to help get the colour right for a lovely miniature we'd decided to have done by Mrs Wood. The session went well and Shandy seemed to enjoy it. But these were to be the very last pictures taken of her on her own as well as with her family. When they were developed, we saw how time had dropped the features and we were reminded of her great age.

Later that day, we decided to take Sally and Samantha for a run in the park. Shandy got excited and was keen to go out with them. We put her lead on and tentatively began a slow walk to the park gate only yards away. Halfway there, Shandy stopped with a rattling noise in her throat and collapsed on the pavement. We knew that it was another attack, this time a bad one. I stayed with her while Douglas took the others back indoors and rang the vet. The next few hours went by like a nightmare; it was rush hour on a Friday evening during August and traffic was heavy. A man passing in his car had seen me with Shandy as she went down, guessed what had happened and came over to help. Douglas was back by this time, bringing the car with him but the passer-by helped lift Shandy into the back seat. She was unconscious by now and her breathing was heavy.

The journey to Croydon at that time of night through the rush hour traffic was agonisingly slow and we arrived in the surgery not a moment too soon. The vet came out, took one look at Shandy, felt her heartbeat; rushed inside for a drip and gave her an injection. She didn't leave the car at all. We brought her home about 7.30. I'd left the dinner in the oven and the vet advised us to leave Shandy in the car for a while. We hurried through dinner and went out to bring Shandy indoors. Alfred and I carried her inside gently on her favourite blanket and laid her in her favourite spot in the lounge, to the right of my armchair by the fireplace. Her breathing was becoming steady and she seemed more relaxed. I knelt beside her, talked to her and stroked her and, although she hadn't regained consciousness, I feel sure she knew where she was and that we were with her. Shandy's breathing became slower and slower and quietly stopped. I looked up and said to Douglas and Alfred, who were still sitting at the table: 'She's gone'. Then I closed her eyes and kissed her and thanked her for being such a lovely girl. The day we'd been dreading had come and we decided to leave Shandy in her favourite spot overnight.

I don't think I cried that night. There was just this dreadful sense of emptiness. We phoned Tom Shilling, who looked after our garden at Tamarisk and he hastily made a big box for Shandy to be buried in next to her little granddaughter, Shelley.

The full impact of her passing didn't really hit me till the following day. We came down in the morning and went to look at her, so peaceful, as if asleep. No one looking at her would have guessed she was no longer with us. It was then the tears began to flow and kept on in waves for the greater part of the day. We set off for the coast that afternoon, putting Shandy, wrapped in the blanket on which she'd been lying, gently in the boot of the car. Gladys came round to see Shandy when we arrived and helped to take her up the twelve steps on the east side of the garden to lay her in the big box. It was approaching tea-time and we hadn't yet had time to dig her resting place, so we left her lying peacefully in the box under a tree nearby till the next day.

It was the town's Carnival evening and Gladys wisely suggested that Alfred and I go with her and her family to watch. It was a lovely sunny evening and Douglas stayed to look after Sally and Samantha indoors. They were very disturbed by Shandy's passing, and had seen her in the lounge to make their own investigations. I feel sure they knew she was no longer with us and accepted it. The Carnival procession helped us through a difficult evening but I only saw about half of it, as my thoughts were with Shandy. I kept telling myself she was a very old lady, that it was right she should go and how lucky we were to have had her for so long. Sunday dawned with a lovely sunny morning. Tom came round, and he and Alfred laid our beloved Shandy to rest next to Shelley. I put a little note in with her, as I have always done with all our dogs, and feel sure that somehow, somewhere, someone will convey that little message to her and, in another dimension, she will understand.

Alfred left on the train and I went to the big church in the town for the evening service, as I often did during our visits. Douglas came with me and I cried more there than I did at home. It was just the right place to say thank-you for her life with us.

In my diary next day I wrote, 'Today starts our first week for the rest of our lives without Shandy in the house.' It was a busy day. I wrote several letters in reply to some lovely cards and letters from 'doggie' friends and others without pets, who'd heard the news. They'd all known Shandy from a pup and knew how close we all were. It was so strange seeing only two 'goldens' around the house.

Before a surround was put round the grave, I asked Douglas to make a little white cross to mark the spot where Shandy had been put to rest. We also put some red roses there from the garden.

A Golden Love Story

We returned to Elmwood the following Thursday. Whilst I couldn't take Samantha to classes that evening as she'd come into season, I had to call in to pick up dog biscuits and everyone wanted to know about Shandy. It was a sad evening. Yet among my other mementoes I have a lovely clipping of Shandy's tail, which I keep in an envelope among my clothes. A year after Shandy's passing we presented a trophy, The Sunspray Cup, in her memory to the Southern Golden Retriever Society, to be awarded to the oldest veteran in the first four at the championship show held in March each year.

Chapter 14
AND THEN THERE WERE NONE

We left with Sally for a visit to our friend Holland in Wales. Sally had to come too as we needed to keep an eye on her and to give her all the medication she needed ourselves. It's a great responsibility for anyone, other than the owner, to care for a sick animal, and the owner's presence always has a more reassuring effect on the pet concerned. She settled down well with us in the peace and quiet of the Welsh countryside and came everywhere with us, including on a trip to Great Orme in Llandudno, where we all had tea in a pretty garden. I felt sure we wouldn't be able to take her in, but everyone welcomed her and she was much admired and spoiled by the owners and customers alike. As it turned out, Samantha had in fact been taken to the vet by my friend Peggy while we were away. She had a tummy upset which had started the day we left, a protest, maybe, at having been left at home. On our return, it still seemed very strange not to see Shandy, and I expected to see her appear from the garden or her favourite spot in the lounge. But it was only mid-August and memories were still very fresh.

Once home, Sally became very disturbed and, after a bad night, she was again at the vet's. Peggy drove us, as my car had suddenly gone on the blink, and Douglas was enjoying one of his rare days out at cricket. I was glad he didn't know before he left for the match, as he would have insisted on taking us and missed a relaxing day watching one of his favourite sports. Sally had developed a limp in the right rear foot and was given an injection. She was very restless all evening, panting and generally out of sorts. Douglas took us back to the vet next morning, as we were both far from happy with her condition and hated to see her in such obvious distress. We were told she had a bad bout of arthritis and when she had a second injection and some tablets, she was certainly much better, almost back to her old self, in fact.

The next day we went to Tamarisk, taking Sally and Samantha with us. I went to the graves as soon as I arrived, and it seemed so strange that we now had two in the garden. Sally enjoyed a trip with us to Bredgar to visit friends with a small holding who sell the most beautiful peas, among other things. When we got home, Sally had a check-up and seemed much improved. One Friday in August 1980, Mrs Wood came to begin the miniature portrait of our beloved pets. Shandy had been gone a month and Mrs Wood would be working solely from the colour picture of her taken the day she passed over. That day Sally was again giving rise for concern. She seemed quite normal during the day, but she became upset by nightfall and very reluctant to settle anywhere, even in her favourite spots. If only we could ask our animals why they behave the way they do.

We had travelled back to Tamarisk and the story there was the same during the day, when Sally seemed normal. But by nightfall, things became a nightmare, as she was so upset. A few days later we returned to Tamarisk again on what was to be Sally's last visit. We took her and Samantha along the front for a good run before setting off for Sydenham and Croydon, to see the vet. We were assured once again that her troubles weren't with her vital functions but were purely muscular and so we accepted this as Sally seemed to be behaving like her old self, going for a good run in the park and again the following day. Douglas came with us and seemed reassured, as was I, that the vet's diagnosis was correct. Sally was as active and normal as any other dog out for a run in the park especially for her age, which was by now thirteen years and four months.

The following Monday started pretty normally, with Sally and Samantha going for a run. But, around noon, Sally went to the bottom of the garden and collapsed by the bird aviary. That finally convinced me the vet was wrong. No one, human or animal, would collapse like that with arthritis. I phoned another vet in Beckenham and asked for a second opinion, but we first had to obtain our own vet's permission for this to be given. The first appointment that afternoon was at four and it was a long, worrying afternoon. Sally had recovered sufficiently to come indoors and lay quietly and we set off with her shortly after. Mr Hawken did a quick examination and asked our permission to do an exploratory operation after looking at her eyes and feeling her tummy. He concluded ominously: 'If I find what I think I'm going to find, I'd like your permission not to wake her up. Let's just hope I'm wrong.' But I knew in my heart he wouldn't be.

He left us with Sally for a few minutes, in which time I said my farewell to her as she sat on the floor of the consulting room. Mr Hawken returned and I handed him the lead and watched Sally walk away with him into the operating theatre and out of our lives.

'We'll go to Croydon now,' said Douglas firmly. This was only a small mission to deliver some things for the PDSA, but it would take our minds of Sally to some extent, and would be better than waiting at home for the phone call telling us how things had gone. We'd given permission for her to be put to sleep should the dreaded cancer be found. But Mr Hawken said he'd telephone anyway.

Alfred took the call and was told that Sally's liver was covered in tumours. He wanted to confirm permission not to bring her round after the operation. Alfred had to give it, there and then, as Sally was, of course, still on the table. But he was quite worried at what he'd done until we returned. It is difficult to say which was the stronger emotion that night - sadness, or anger at Sally being allowed to suffer for as long as she did because of a wrong diagnosis.

It is always easy to condemn, but I think we're entitled to expect more care and attention for animals, especially since they cannot talk to us about their symptoms. We felt badly let down and promptly changed our vet. We continue to use Mr Hawken and Mr Jefferies at Beckenham to this day.

Douglas and I collected Sally's body from Mr Hawken next day, wrapped in her big white blanket. He'd cut a piece of fur from her lovely tail, at my request, as a keepsake. We then set off for her last journey to Tamarisk. Alfred came with us to help with her burial and Tom made a really nice box at short notice. The next morning was a lovely sunny one and we laid our darling Sally to rest in the sunshine next to her sweet mother, Shandy. I put the bundle of old socks she always brought singing to me whenever I'd been away and a note to her in the box. I placed some roses on the same mound of earth. Douglas comforted me, saying, 'Sally is alive in the other world and enjoying being with Shandy and Shelley.' That's what we believe and that's what we must stick to. Sally had been reunited with darling Shandy after six short weeks. Lucky girl.

* * * *

The following day seemed unreal, and the house strangely quiet. Douglas had gone to Burgess Hill to help a friend out with a business. He was in town all evening at a meeting and I was at the dog club with Samantha, who was very quiet indeed. I stayed on the whole evening to take the money as the treasurer was ill, so we were well occupied, which helped to soften our loss a little. Mrs Wood called to do some more work on the miniature with Samantha and it was coming along nicely. The following day, I helped Alfred get ready for his holiday. He was particularly sad and upset at Sally's passing as he somehow regarded her as his dog, having chosen her name. I went to a dog training club committee meeting that evening, where there was much talk of Sally.

We went to Tamarisk with Samantha that weekend and I sat in the garden near those sad little resting places all afternoon and, in the evening, we both took Samantha along the front for a run. How strange it seemed, after having three dogs to run for so long. Sally had been gone a week and, despite all Douglas's efforts to keep me occupied, with a trip to Hastings with Samantha and dinner back at Fordwich, nothing seemed to overcome the feeling of loss and emptiness. It was at least two weeks before this began to lift and I felt brighter again. After all, we still had Samantha and, even if she tended to be a bit of a loner, she loved and needed us and we loved and needed her in return.

* * * *

Our annual dog show for the PDSA came round again. Alfred took Samantha, who was awarded a VHC veteran and certainly seemed to be enjoying life. She must have missed the company of her mother and grandmother, of course. Although they rarely played together, they always ran together and she must have gained moral support from their presence. Samantha had done a few laps at the dog show and sponsored walk, and was joined by Shandy's daughter, Mandy, her daughter, Heidi, and another 'golden' friend, Cash, (the banking family's dog) who'd come up from Buckinghamshire to walk for us. I must say we were very proud of them all.

We went to Elizabeth Wood's house on 7 October to see the miniature of our three beloved pets finished and I was truly delighted with the result. It was so good of them all and Sally in profile was perhaps the best. We had a most agreeable morning, drinking coffee and chatting to Mrs Wood and her husband. All that remained was the frame for the picture and then I would be wearing it.

We had some difficulty with Samantha on our next visit to Tamarisk. She kept going to Sally's resting place and digging. We'd never before experienced this and it was rather upsetting. Douglas thought she was somehow getting the scent of her mother, even though the grave was quite deep and well covered, and it was sometime before we managed to persuade her to stop. Our annual dog club dance took place in November, with awards being presented. When the new novice shield was handed out, I found it was dedicated to Shandy and Sally, a lovely and most touching surprise. The snow arrived early that winter and, on our next weekend in Kent, the wind and snow stopped me from going to the top garden to the girls' resting place. But the conditions were better just before we returned and I was able to put flowers from our conservatory on Shandy's plot for her birthday. We called on Mrs Wood to collect the miniature, now completed and in its case. I must say, it looked even more beautiful. Douglas bought me a gold chain and I wear this lovely picture every day and everywhere.

Chapter 15
THE LAST GOODBYE

Three days before Christmas, I'd been out shopping for quite a long time and collecting alterations from the dressmaker. On entering the cottage in the late afternoon, I realised how much I missed our two old ladies, Shandy with her gentle welcome and Sally bounding up to me singing and carrying her bundle of old socks as a welcome home gift. It was our first Christmas in Elmwood without Shandy. She'd spent fifteen of them with us and Sally twelve, and it was very strange to have Samantha there on her own. But we were grateful for her company, even though we had a gathering of friends due both before and during Christmas.

I always enjoyed the homely scene with the dogs around our feet and, more often than not, on our laps by the log fire. On Christmas Day, Samantha walked with me to visit an old lady friend who was 95 and delighted to see her. The lady in question is still going strong at 103 and still asks after our present dogs whenever I see her.

* * * *

Samantha celebrated her eleventh birthday in February and I was working at a PDSA sale that day and thinking how happy we were eleven years ago, with our beloved Sally and her new-born family. We left for a holiday in Egypt in May. It was the worst holiday we'd ever had. Douglas flew out with a bad cough which was made much worse with the extreme heat and dust. The only good thing about the trip was seeing the splendid work of the PDSA hospital in Cairo and a visit to the *son et lumière* at Giza, which was most impressive. We became friendly with two charming sisters from London, named Jessica and Miriam, who agreed with us that the whole place showed a lamentable lack of cleanliness and hygiene. Our great joke was the sawdust biscuits which were served at tea-time. I thought of Samantha, who might have appreciated those biscuits, back home in the cool of England and I longed to be back there.

We arrived home on 16 May and it was beautiful to feel the chilly breeze after 120 degrees in the shade at Luxor. The greeting from Samantha was well worth the wait. Oh, it was lovely to be home! Douglas still had his bad cough and tummy upset but soon responded to treatment and life once again returned to normal.

The end of May saw a change in venue for Beckenham Dog Training Club and we had our last training session in Leonard Road, Sydenham, where we'd been for nine years. It was a sad evening for me, as the old church hall was the last one Shandy visited the evening before she passed away and one all the girls had done well in. It was, however, a bitterly cold place, having no heating of any kind, and I think our club and committee members should have had medals for endurance. A block of flats now

A Golden Love Story

stands on the site. At the new venue, Samantha came third in a veteran class in June. Alfred took her as I was feeling unwell with sickness and general lethargy. My left hand became painful and swollen, and blood tests revealed that I'd picked up a bug which affects the liver whilst in Egypt, where it is quite common. I had to go on medication and a completely fat-free diet and I can still hear our doctor calling along the hall as we left: 'Bournemouth next year!' Our Sally's puppy Rusty, along with John and Ethel, acquired a new playmate called Sandy, another 'golden' puppy. We called to see them and were delighted to learn how well Rusty had received the newcomer. The baby seemed to give the old fellow a new lease of life, as so often happens. Rejection can take place but, on the whole, is rare, especially if care is taken to introduce the newcomer carefully and the old dog gets every bit as much fuss and attention as the pup.

My aunt who lived in Eastbourne, died suddenly and unexpectedly that month. As Douglas and I were both executors of her will, we had to go down to her bungalow and attend to her affairs. Each time we went, including the day of the funeral, we took Samantha with us to run free both indoors and in the garden. I smiled as I watched her running around, sniffing, exploring and generally making herself at home. My aunt didn't dislike animals, but she was so intensely house proud that my uncle, who would have loved a dog, was not allowed to keep one. I tried to imagine the look on my aunt's face at the scene. We went to Chepstow races to watch Eagle's Quest run and were able to have a reunion in Newport with our Bobby and Salvador. They reminded me so much of Shandy and Sally. I took them for a walk before dinner and we took some photographs in the garden but the weather and light were poor, so they were unsuccessful, which was sad as it was the last time we were to see the dogs. Mrs Wagstaff brought Koppa to see us a few weeks later. He was a lovely boy and so clearly adored. But that, too, was to be our last meeting.

* * * *

Samantha had an operation to remove a wart which had grown quite large on her side and had started to bleed. I often wonder if it was an external forerunner of what was to come, but all went well and the stitches were removed after two weeks. Christmas 1981, was white and everything Samantha loved. How she adored the snow, throwing herself into it and rolling over and over. It was bitterly cold just before the snow came a few days before Christmas and, when it did fall, it was thicker than I'd remembered for a long time. Looking back, I'm so glad it was, as it was the last snowy festive season Samantha would ever enjoy.

I celebrated my fiftieth birthday on 1 February 1982, a real milestone for me. I was sad that all the ladies weren't alive to share it. Samantha was still with us and spent a fairly normal day running in the park and being looked after when I went for a celebration lunch with Douglas and Gladys before a party at home for 36 friends. That particular event didn't meet with Samantha approval. There were far too many people around in her favourite relaxing place in the lounge, so she took herself upstairs to rest on Alfred's bed for the duration of the party. She had one of her tummy upsets about a week later, which appeared to right itself on a diet of boiled rice and Slippery Elm

The Last Goodbye

food. Samantha had her twelfth birthday on 28 February and her special treat was a dinner of wild rabbit and a visit to our old lady friend in College Road.

* * * *

Like Sally, Samantha had to have her anal glands cleaned and, when we went to the vet, he discovered she was having a phantom pregnancy at what, in a human, would be 84 years old!

Our old lady, Mandy, together with Heidi, came with me and Milly to help collect for the PDSA at a charity fair in Bromley. The presence of dogs or any other animals always attracted a good deal of attention and fuss, as well as more money for a worthy cause.

1982 was not my best year and small changes on the domestic scene weren't at all to my liking. To begin with, I always keep my cars too long and hate having to part with them. My red Opel Ascona, which had been with me for ten years, left to go to friends in Whitstable. The fact that it was going to Val and Chris made me feel a little better, but, as they drove away in it, I kept thinking back to all the fun I'd had in that car and the many journeys the girls and I had made together. I would see it often in Whitstable in future years, unexpectedly in a car park or a side street, still in good condition for its age. And, in my mind's eye, I could still see our beloved girls sitting in the back and looking out of the window once more.

One day in May Samantha became very sick and we took her to the vet, who found she was suffering from a high degree of anaemia. He gave her an injection and loads of tablets and, for the first time, I began to feel worried about her, remembering that Sally had gone down the same way. We took Samantha back to the vet's for a blood test on her liver and the vet phoned to say the test was all clear. Within a few days, she was eating well and was much more like her old self, so we began to relax. But soon we were back to square one. We had taken her with us to Tamarisk for a few days and she became unwell with the same symptoms. Back we went to our vet in Whitstable, who kept her in overnight on a drip and took more blood tests.

We called on the vet the following day and Samantha was much brighter. We collected her at 5 pm and she tucked into a good meal of rice pudding and scrambled egg. But, on her return to Sydenham, she was to have an operation to determine the cause of the trouble which was now disturbingly frequent. We came back the next morning, leaving Tamarisk early and getting Samantha to Mr Jefferies before noon. I felt very worried about her, and deep down I knew what the trouble was. It was just a matter of having it confirmed. We left Samantha to have her operation and took a drive round. Then we went home to sit in the shade of the garden and watch the clock until it was time to telephone the vet for news. The operation had gone well and Samantha's spleen had been removed, covered in tumours. She stayed in the surgery for two nights and the following morning I phoned to ask about her condition. She had had a good night and was recovering well, they told me. We went about our business, having lunch with friends in London and going to watch the Derby. I had a check-up with our own doctor and counted many flag day tins for the PDSA with guild friends.

We decided to put our planned holiday in Devon back two weeks to give Samantha time to recover fully under our supervision. We collected her and she

A Golden Love Story

seemed bright, tucking into a good meal indoors. But her many stitches bothered her so I decided to sleep downstairs with her. Samantha had a surprisingly good night and I rang the vet to tell him we were pleased with her progress. She had lots of tablets to take at various times and Alfred took a turn at staying down with her. 5 June was the hottest day of the year, 86 degrees, and keeping Samantha cool and stopping her from licking her stitches was quite a problem. I left her with Douglas for the afternoon and went to Blackheath for a demonstration by Beckenham Dog Training Club and to see a small exemption show for the PDSA, in conjunction with the 'Village Fayre'. I was thankful to get home to Douglas and Samantha and out of the heat. The heat seemed to fade out into more reasonable temperatures and we took Samantha to Mr Jefferies for her check-up. He was pleased with her progress but I still slept downstairs with her every other night, until her stitches were removed.

Samantha was still very much up and down despite the good report. We took her with us the following week, after her check-up, to Tamarisk but she seemed far from well. She rested in the garden and later on seemed fine again, enjoying a good meal. The next day was much cooler and Samantha seemed happy, bounding around the garden and coming on short walks.

The improvement continued until Saturday, when she was most unwell after her dinner. We called the vet, who reassured us that there was nothing to worry about. He seemed to be right as, sure enough, later that evening she was fine. Her stitches came out and she was as good as gold and, seemingly, as good as new. She was more relaxed with the stitches gone and decided to walk upstairs for the first time in two weeks. Soon we were back to our two short runs a day in the park.

* * * *

I took Samantha to the park early on 21 June and then left with Douglas for our postponed holiday. I wasn't entirely relaxed about Samantha, but she was well cared for at home and we were only a telephone call away. I was glad we weren't going abroad. We had a lovely week and I called home at regular intervals to be told all was well. But I still had this feeling of unease, off and on. We stayed an extra couple of days with friends in Dorset on the way home and arrived back in Elmwood to be greeted by a boisterous Samantha, looking good. My fears melted and my spirits rose!

Another check-up the following day seemed to confirm that all was in order and the vet was well pleased, so we took off for a visit to our friends in North Wales and enjoyed a lovely week with them, visiting the races at Chester and browsing around craft fairs. Douglas loved the races and the whole atmosphere appealed to him, mostly because he was out in the open air among the horses but not a great crowd of people.

Telephone calls home reassured us that everything was fine. We were entertained that week by the brother of Dick Francis at the Francis Stables in Malpas and Douglas was in his element, touring the stables and learning the history of every horse there. We were lucky enough to go because the friends on whose farm we were staying had a horse in training there at the time.

Back home, I took Samantha for her usual run in the park. It was good to see her and I was grateful she looked so bright and happy and to know she'd been well cared for while we'd been away.

Life went on rather erratically for a few weeks, with Samantha fine for a few days and then suffering bouts of sickness off and on. Yet, even during these spells, she would be as bright as a button, running around all her favourite spots and eating well.

* * * *

Samantha didn't seem well again on 7 August. She was perky enough but something told me all was not right with her. The vet prescribed iron tablets and again took her in overnight. It was very hot and I lay out in the garden on what should have been a relaxing summer's afternoon. But it was impossible to unwind when I was so worried about Samantha, knowing she was in the best place but wishing she could be at home with us and her bonny self again.

I was on edge again next day and kept as busy as possible till tea-time when we went back to the vets to collect Samantha. We had a long chat to Mr Jefferies, who had sent more blood samples away, but he wasn't hopeful of a good report. In fact, the overall picture was decidedly bad. Samantha wasn't eating and we kept her very quiet.

It was a tearful evening, helped only by the fact that I'd left Samantha with Douglas and gone to the dog club, where I took two classes as it was holiday time and there was a shortage of instructors. I hurried home as soon as my second class was over to be with Samantha. Alfred slept down with her for the night. The next morning, Samantha wouldn't touch her breakfast but was keen to go to the park, a treat she still enjoyed. To see her running around out there, you would never have guessed she was so ill. She did eat some liver and beef mixed up later in the day, and then rested quietly at our feet. No one was very hungry at dinner. Worry and grief somehow diminish the appetite. I slept down with Samantha that night and, next morning, the story was the same. Samantha was bright and eating. Douglas and I took her for a run in the park, which she really loved. The results from the laboratory on her blood test were as expected: her liver wasn't making red blood cells and we were to continue with iron tablets.

Samantha and I went to Crystal Palace grounds for what was to be her very last run. She had a lovely half-hour walking, sitting, rolling and sniffing and, as we were about to come up the slope which led to the gate, she turned, sat down in the middle of the path and took a long look all round, as if she knew she was seeing it all for the last time. When she was ready, she got up and we slowly strolled home. Douglas and I had booked to go to Chartwell that evening to see the *son et lumière* of Churchill's life. Samantha was contented after her two trips out and her meal, which she seemed to enjoy. Alfred stayed with her for the evening but, had I realised this was to be her last on earth, I would have suggested we call off our trip and Douglas would have been the first to agree.

We went to dinner first, as planned, and on to Chartwell. It should have been a lovely summer's evening out in the garden at Churchill's old house. But my mind, like Douglas's, was very much on Samantha. As I watched that performance, I knew beyond any doubt, we would lose her very soon.

I opted to stay down with her for the night and the following morning, before eight, we knew the dreaded day had arrived. Samantha couldn't take food or water and was dreadfully sick. We couldn't allow this situation to go on any longer, the cancer

A Golden Love Story

had now destroyed her liver and it would have been cruel to try to keep her alive. It was a tearful Douglas who'd broken the news to me that morning. He'd got up early and come down to feed Samantha and was greatly upset to see the sorry state she was in. So we agreed I should call Mr Jefferies. Samantha, meanwhile, lay in the cool of the hall until mid-morning then she suddenly got up and walked through the kitchen and breakfast room, just beyond where she was born, a little under twelve and a half years before. She went out into the garden, turned right and rested in the sun.

Soon Mr Jefferies arrived with a sweet nurse to assist him in the sad task. There was no rush. I had all the time I wanted and needed to say my farewell to Samantha out there in the sunshine on this summer's day. At last I told Mr Jefferies we were ready. He was all prepared and Douglas and I stayed with Samantha, with the vet on one side, and the nurse on the other.

So, after many kisses and strokes and words of gratitude for her life, it was at 11.35 on Thursday 12 August 1982, that our Samantha, last of our beloved doggy family left us. I turned to Mr Jefferies as soon as the injection had been given and asked: 'Has she gone?' to which he replied, 'Yes, her soul has gone.' They were at the same time the saddest and most comforting words he could have spoken. I managed to say: 'She's a lucky girl to be out of her suffering and reunited with her family.' What more could any creature, human or otherwise, wish for? Only her human family weren't yet where she'd gone. But, all in good time.

The rest of the day was unreal. Mr Jefferies gently carried Samantha out to his car and back to the surgery, where he would keep her until she was ready for burial. The afternoon was very hot and I knew I couldn't stay indoors for the entire day. Within an hour of Samantha being taken away from Elmwood, I went for a walk alone for the first time since 1965 in Crystal Palace grounds. There were no tears on this walk, only a feeling of numbness and disbelief that our little family had all at last gone and the feeling of nostalgia that always sets in on these occasions. In my mind's eye, I could see them all running ahead, spreading out before me in all directions as they went, and joining up again in the distance for a lovely romp. Then they'd respond to my call to come and have their leads put on before we reached the gate and walked the last few yards home.

We collected Samantha from the surgery and made the sad drive to Kent for her burial. We discovered her resting place had been dug up by our gardener, not in line with the others, as I'd asked, but on the side next to her grandmother and overlooking Sally and Shelley. At first, this was a disappointment to me, as I had visions of the resting places coming down in a line in the order in which they'd passed on. But, after much deliberation, it was decided to leave the place as it was. We took the black plastic covering bag off to reveal what could have been a peaceful Samantha, curled around in a dog's natural sleeping position. She was wrapped in her green woollen blanket and looked so much at peace, I really felt glad for her.

Alfred and I gently lifted her into the box, so beautifully made in such a short time. As with the others, a little note went in, with the words known only to God, the dog concerned and myself, a cosy secret between Our Creator, the girls and me. And so, there they were, all together again and, as Samantha had always sat behind her

The Last Goodbye

relations in life, so she was again in the same position. Was it just a strange coincidence that Tom had dug her last resting place in that particular spot?

* * * *

One day, in the not too distant future, I like to think that I shall fall asleep, in an armchair or in my bed and, when I wake up, find myself in a quiet, leafy country lane, with beautiful fields, trees and woods, streams, flowers, birds and other wildlife around. And, of course, the sun will be shining. As I walk over the brow of the hill, everyone near and dear to me in life will come towards me and, bounding along in front of them, full of fitness, youth and joy, will be our beloved girls joined undoubtedly by then by Simon, Sarah and Selina, our present dogs. The last tears I shall ever shed will be tears of joy at that wonderful reunion. Meanwhile, they all lie at peace in our lovely garden overlooking the sea, enshrined forever in our love.

He has but turned a corner
He is not dead this friend
But in the path we mortals tread
Got some few trifling steps ahead
And nearer to the end
So that you too, once past the bend
Shall meet again.
He loiters with a backward smile
Till you can overtake.

ROBERT LOUIS STEPHENSON

Also from The Lutterworth Press:

First Aid and Health Care for Dogs
Charles Bell

As a practising vet, Charles Bell has had the opportunity to meet many distressed pet owners who would love to know what they can do to help when their pet meets with an accident. In response to this demand he has written a comprehensive manual for dog owners, covering not only basic first aid in a variety of emergency situations, but also giving sound practical advice on general welfare and health care. He includes tips on travelling with dogs, dental care, dieting and exercise as well as a guide to rehabilitation after an operation or accident. This book is in no way intended to replace the vital function of the veterinary surgeon but is intended to provide a bridge between the dog owner and the vet, enabling the owner to cope with any eventuality until the vet can be contacted.

Charles Bell graduated from Cambridge University with a degree in Veterinary Medicine and has practised for a number of years both in the United States and in England, specialising in small animals.

Also from The Lutterworth Press:

My Puppy is Born
Joanna Cole

Especially for young dog lovers, this interesting and informative book tackles a subject which can sometimes be frightening and upsetting for children in a sensitive and uncomplicated way. It should prove an invaluable aid to parents and teachers of four to six year-olds who are wondering how to approach the subject of birth. Colourful, step-by-step photographs trace the birth and early life of Dolly, a Norfolk terrier puppy up to the day she is ready to enter her new home.

Joanna Cole is the successful author of many award-winning books for children such as *A Gift from St Francis*, *Anna Banana*, and the *Magic School Bus* series.